Student's Guide to
Calculus
by J. Marsden and A. Weinstein
Volume III

Frederick H. Soon

Student's Guide to
Calculus
by J. Marsden and A. Weinstein
Volume III

With 186 Illustrations

Springer

Mathematics Subject Classification (1991): 26-01

Library of Congress Cataloging in Publication Data
Soon, Frederick H.
  Student's guide to Calculus by J. Marsden and
A. Weinstein, volume III.
  Guide to Marsden and Weinstein's Calculus III,
  1. Calculus.   I. Marsden, Jerrold E.   Calculus III.
II. Title.
QA303.S7745   1986   515   86-6572

Printed and bound by Electronic Printing Inc., Plainview, NY.
Printed in the United States of America.

9 8 7 6 5 4 3

ISBN 0-387-96348-0 Springer-Verlag New York Berlin Heidelberg
ISBN 3-540-96348-0 Springer-Verlag Berlin Heidelberg New York SPIN 10730453

Dedicated to:

Henry, Ora, Dennis, and Debbie

This Student Guide is exceptional, maybe even unique, among such guides in that its author, Fred Soon, was actually a student user of the textbook during one of the years we were writing and debugging the book. (He was one of the best students that year, by the way.) Because of his background, Fred has taken, in the Guide, the point of view of an experienced student tutor helping you to learn calculus. While we do not always think Fred's jokes are as funny as he does, we appreciate his enthusiasm and his desire to enter into communication with his readers; since we nearly always agree with the mathematical judgements he has made in explaining the material, we believe that this Guide can serve you as a valuable supplement to our text.

To get maximum benefit from this Guide, you should begin by spending a few moments to acquaint yourself with its structure. Once you get started in the course, take advantage of the many opportunities which the text and Student Guide together provide for learning calculus in the only way that any mathematical subject can truly be mastered — through attempting to solve problems on your own. As you read the text, try doing each example and exercise yourself before reading the solution; do the same with the quiz problems provided by Fred.

Fred Soon knows our textbook better than anyone with the (possible) exception of ourselves, having spent hundreds of hours over the past ten years assisting us with its creation and proofreading. We have enjoyed our association with him over this period, and we hope now that you, too, will benefit from his efforts.

Jerry Marsden

Alan Weinstein

HOW TO USE THIS BOOK

As the title implies, this book is intended to guide the student's study of calculus. Realizing that calculus is not the only class on the college student's curriculum, my objective in writing this book is to maximize understanding with a minimum of time and effort.

For each new section of the text, this student guide contains sections entitled Prerequisites, Prerequisite Quiz, Goals, Study Hints, Solutions to Every Other Odd Exercise, Section Quiz, Answers to Prerequisite Quiz, and Answers to Section Quiz. For each review section, I have included the solutions to every other odd exercise and a chapter test with solutions.

A list of prerequisites, if any, is followed by a short quiz to help you decide if you're ready to continue. If some prerequisite seems vague to you, review material can be found in the section or chapter of the text listed after each prerequisite. If you have any difficulty with the simple prerequisite quizzes, you may wish to review.

As you study, keep the goals in mind. They may be used as guidelines and should help you to grasp the most important points.

The study hints are provided to help you use your time efficiently. Comments have been offered to topics in the order in which they appear in the text. I have tried to point out what is worth memorizing and what isn't. If time permits, it is advisable to learn the derivations of formulas rather than just memorizing them. You will find that the course will be more meaningful to you and that critical parts of a formula can be recalled even under the stress of an exam. Other aspects of the study hints include clarification of text material and "tricks" which will aid you in solving the exercises. Finally, please be aware that your instructor may choose to emphasize topics which I have considered less important.

Detailed solutions to every other odd exercise, i.e., 1,5,9, etc. are provided as a study aid. Some students may find it profitable to try the exercises first and then compare the method employed in this book. Since the authors of the text wrote most of the exercises in pairs, the answers in this book may also be used as a guide to solving the corresponding even exercises. In order to save space, fractions have been written on one line, so be careful about your interpretations. Thus, $1/x + y$ means y plus $1/x$, whereas $1/(x + y)$ means the reciprocal of $x + y$. Transcendental functions such as cos, sin, ln, etc. take precedence over division, so $\cos ax/b$ means take the cosine of ax and then divide by b , whereas $\cos (ax/b)$ has an unambiguous meaning. $\ln a/2$ means half of $\ln a$, not the natural logarithm of $a/2$. Also, everything in the term after the slash is in the denominator, so $1/2\int xdx + 1$ means add 1 to the reciprocal of $2\int xdx$. It does not mean add 1 to half of the integral. The latter would be denoted $(1/2)\int xdx + 1$.

Section quizzes are included for you to evaluate your mastery of the material. Some of the questions are intended to be tricky, so do not be discouraged if you miss a few of them. The answers to these "hard" questions should add to your knowledge and prepare you for your exams. Since most students seem to fear word problems, each quiz contains at least one word problem to help you gain familiarity with this type of question.

Finally, answers have been provided to both the prerequisite and section quizzes. If you don't understand how to arrive at any of the answers, be sure to ask your instructor.

In the review sections, I have written more questions and answers which may appear on a typical test. These may be used along with the section quizzes to help you study for your tests.

Since _Calculus_ was intended for a three semester course, I have also included three-hour comprehensive exams at the end of Chapters 3, 6, 9, 12, 15, and 18. These should help you prepare for your midterms and final examinations. Best of luck with all of your studies.

ACKNOWLEDGEMENTS

Several individuals need to be thanked for helping to produce this book. I am most grateful to Jerrold Marsden and Alan Weinstein for providing the first edition of _Calculus_ from which I, as a student, learned about derivatives and integrals. Also, I am deeply appreciative for their advice and expertise which they offered during the prepartion of this book. Invaluable aid and knowledgeable reviewing were provided by my primary assistants: Stephen Hook, Frederick Daniels, and Karen Pao. Teresa Ling should be recognized for laying the groundwork with the first edition of the student guide. Finally, my gratitude goes to my father, Henry, who did the artwork; to Charles Olver and Betty Hsi, my proofreaders; to Sean Bates for his contributions; and to Ruth Edmonds and Esther Zack, whose typing made this publication a reality.

Frederick H. Soon
Berkeley, California

CONTENTS

CHAPTER 17 -- MULTIPLE INTEGRATION

CHAPTER 18 -- VECTOR ANALYSIS

13.1 Vectors in the Plane

PREREQUISITES

1. There are no special prerequisites for this section other than basic
   high school algebra.

GOALS

1. Be able to perform scalar multiplication, addition, and subtraction

   with vectors.

STUDY HINTS

1. <u>Adding ordered pairs</u>. The sum of two ordered pairs is another ordered
   pair. The procedure is simply to sum the corresponding components
   like this: $(x_1,y_1) + (x_2,y_2) = (x_1 + x_2, y_1 + y_2)$.

2. <u>Scalar multiplication</u>. A scalar is simply a real number, as in Section
   R.1. Scalar multiplication involves a number times an ordered pair, represented by $r(x,y) = (rx,ry)$. This is opposed to a vector times a vector, which will be discussed later in the chapter.

3. <u>Uniqueness of solution</u>. Example 4 illustrates a general principle: n
   equations in m unknowns will <u>not</u> have a unique solution if n < m .

4.   Vectors defined. A vector has both length (magnitude) and direction.

 A scalar does not have direction. Two vectors
are equal if and only if they both have the
same length and the same direction. Pictorially,
they do not need to originate from the same
starting point. The vectors at the left are equal.

5.   Notation. In the text, vectors are usually denoted with boldface letters.
In this student guide, vectors are usually denoted with an underlined
letter. Sometimes, the vector going from point P to point Q is written
as $\overrightarrow{PQ}$. A vector may also be represented by its components (a,b) . Since
(a,b) may denote either a vector or a point, you should be careful with
your interpretations. Your instructor or other textbooks may use other
notations such as a squiggly line (~) underneath a letter or a circumflex (^)
over a letter to represent a vector. All of these notations are different,
but they all denote a vector.

6.   Vector algebra. Addition and scalar multiplication are analogous to the
same operations with ordered pairs. Geometrically, $\underline{u} + \underline{v}$ is the
diagonal of the parallelogram spanned by $\underline{u}$ and $\underline{v}$. See Fig. 13.1.12.
To multiply a vector by r , extend the length of the vector by a factor
of r . Reverse the direction if r < 0 .

7.   Vector subtraction. If a sketch like Fig. 13.1.13 is desired, one may
forget which direction the difference is pointing. Here is how to
remember. If you want $\underline{v} - \underline{w}$, then you should be able to add $\underline{w}$ to
obtain $\underline{v}$. If you don't get $\underline{v}$, the direction is not correct.

SOLUTIONS TO EVERY OTHER ODD EXERCISE

1.   For addition of ordered pairs, the sum of (x_1, y_1) and (x_2, y_2) is
$(x_1 + x_2, y_1 + y_2)$. Thus, (1,2) + (3,7) = (1 + 3, 2 + 7) = (4,9) .

5. By addition of ordered pairs, $(1,2) + (0,y) = (1, 2 + y)$. For this
    to be equal to $(1,3)$, we need $1 = 1$ and $2 + y = 3$. Thus, $y = 1$.

9. Using scalar multiplication, $2(1,b) + (b,4) = (2,2b) + (b,4)$. Then,
    addition of ordered pairs yields $(2 + b, 2b + 4)$. Therefore, we need
    $2 + b = 3$ and $2b + 4 = 4$. For the first case, $b = 1$; and $b = 0$
    for the second. Since b cannot be both 1 and 0 simultaneously,
    there is no solution.

13. Using scalar multiplication, $a(1,1) + b(1,-1)$ becomes $(a,a) + (b,-b)$.
    Then addition yields $(a + b, a - b) = (3,5)$. Solving $a + b = 3$ and
    $a - b = 5$ simultaneously, we get $a = 4$ and $b = -1$.

17. By addition of ordered pairs, $A + 0 = (x_1, y_1) + (0,0) = (x_1, y_1)$,
    which again is A .

21. By scalar multiplication, the left-hand side is $a(bA) = a(b(x_1, y_1)) =$
    $a(bx_1, by_1) = (abx_1, aby_1)$. Similarly, $(ab)A = (ab)(x_1, y_1) =$
    (abx_1, aby_1) .

25. (a) We represent the molecule $S_x O_y$ (x atoms of sulfur and y atoms
        of oxygen) by the ordered pair (x,y) . Then the chemical equation
        is equivalent to $k(1,3) + \ell(2,0) = m(1,2)$.

    (b) From part (a), we have $k + 2\ell = m$ and $3k + 0 = 2m$.

    (c) From the second equation, we have $m = (3/2)k$. In order to make
        k and m integers, let $k = 2a$, so $m = 3a$. Substituting
        these values of k and m into the first equation gives $2a + 2\ell =$
        $3a$, i.e., $\ell = a/2$. The smallest integer a , which would make
        ℓ an integer is $a = 2$. Thus, $k = 4$, $m = 6$, and $\ell = 1$.

29. (a)

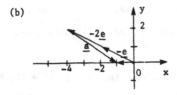

Using component notation, we have $\underline{c} + \underline{d} = (3,2) + (3,0) = (6,2)$. The vector $\underline{c} + \underline{d}$ is drawn analogous to Fig. 13.1.5.

(b)

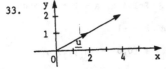

Component notation yields $-2\underline{e} + \underline{a} = -2(2,-1) + (3,-2) = (-4,2) + (3,-2) = (-1,0)$. $-2\underline{e}$ is constructed similar to the method in Fig. 13.1.8.

33.

First, draw $\underline{u}$. Then, extend the vector twice as far in the same direction to get $2\underline{u}$. The vector $2\underline{u}$ has components $(4,2)$.

37. (a)

(b) $\underline{v} = Q - P = (3,3) - (2,1) = (1,2)$; $\underline{w} = R - Q = (4,1) - (3,3) = (1,-2)$; $\underline{u} = P - R = (2,1) - (4,1) = (-2,0)$.

(c) $\underline{v} + \underline{w} + \underline{u} = (1,2) + (1,-2) + (-2,0) = (0,0) = \underline{0}$, the zero vector.

41. (a) Let $\underline{v} = (0,0)$, $\underline{w} = (1,1)$, and $s = 0$, then $r(0,0) + 0(1,1) = \underline{0}$ for any non-zero r . Thus, the definition of linear dependence is satisfied by $(0,0)$ and $(1,1)$.

(b) Intuitively, two vectors are parallel if and only if they have the same direction. Thus, if $\underline{v}$ and $\underline{w}$ are parallel, we can write $\underline{v} = -(s/r)\underline{w}$. This only makes sense if $r \neq 0$, so we can rearrange the equation to get $r\underline{v} + s\underline{w} = \underline{0}$. Therefore, parallel vectors are linearly dependent. Now, if the vectors are linearly dependent, we can go backwards from $r\underline{v} + s\underline{w} = \underline{0}$ to get $\underline{v} = -(s/r)\underline{w}$. Thus, linearly dependent vectors are parallel.

41. (c) If $a = 0$, then $c = 0$ and $ad = bc = 0$. If $b = 0$, then

 $d = 0$ and $ad = bc = 0$. If $a \neq 0$ and $b \neq 0$, then $r\underline{v} +$

 $s\underline{w} = 0$ if and only if $(ra,rb) + (sc,sd) = \underline{0}$, if and only if

 $ra + sc = 0$ and $rb + sd = 0$. Solve each of these equations

for r and equate the results to get $-sc/a = -sd/b$. Cancel

 $-s$ and cross-multiply to get $ad = bc$ if and only if $r\underline{v} + s\underline{w} = \underline{0}$.

(d) Let $\underline{u} = (a,b)$, $\underline{v} = (c,d)$, and $\underline{w} = (e,f)$. Since $\underline{v}$ and $\underline{w}$

are linearly independent, $cf \neq de$ from part (c). Hence $c \neq 0$

or $d \neq 0$. Then $x\underline{v} + y\underline{w} = \underline{u}$ if and only if $(cx,dx) + (ey,fy) =$

 (a,b) , if and only if $cx + ey = a$ and $dx + fy = b$, if and

only if $x = (a - ey)/c$ and $x = (b - fy)/d$ (and $c \neq 0$, $d \neq 0$).

Hence, $(a - ey)d = (b - fy)c$ if and only if $ad - dey = bc -$

 cfy if and only if $ad - bc = (de - cf)y$ if and only if $y =$

 $(ad - bc)/(de - cf)$. Substitute this for y in the equation for x :

 $x = [a - e(ad - bc)/(de - cf)]/c = [a(de - cf) - e(ad - bc)]/$

 $(c(de - cf)) = (ade - acf - ead + bce)/(c(de - cf)) = (be - af)/$

 $(de - cf)$. If $c = 0$ and $d \neq 0$, then $y = a/e = (ad - bc)/$

 $(de - cf)$ and $x = b - f(a/e)/d = (be - af)/(de - cf)$. Similarly,

if $d = 0$ and $c \neq 0$.

Hence, for any $\underline{u}$, $\underline{v}$, and $\underline{w}$, we can specify x and y

in terms of the coordinates of $\underline{u}$, $\underline{v}$, and $\underline{w}$.

SECTION QUIZ

1. On the xy-plane, draw a vector $\overrightarrow{PQ}$ and another vector $\overrightarrow{PR}$.
   (a) What is $\overrightarrow{PQ} - \overrightarrow{PR}$?
   (b) Sketch $\overrightarrow{PR} + \overrightarrow{PQ}$.

2. If the following vector expression is defined, simplify it. If not, explain why it is not defined: $3[(1,2 + 3) + 4(2,5) - (-(2 - 1),-3)]$.

3. What is the difference between a vector and a scalar?

4. A chihuahua was minding his own business at $(1,0)$ when the mailman came to make his delivery. The little dog playfully followed the mailman down the street to $(0,0)$ and then around the corner to $(0,1)$. At that point, the irritable postman said, "Go away, dog! You bother me!" and kicked the chihuahua. In retaliation, the chihuahua bit the postman's leg.

    (a) What vector described the displacement of the dog along the first block?

    (b) What vector described the dog's total displacement?

    Simultaneously, across town, a great dane followed a fearful mailman from $(98,25)$ to $(98,27)$, and then across a vacant lot to $(97,26)$.

    (c) Does the same vector describe the chihuahua's and the great dane's displacement? Explain your answer.[*]

ANSWERS TO SECTION QUIZ

1. (a) $\overrightarrow{RQ}$

    (b)

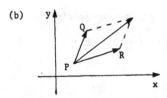

[*] Dear Reader: I realize that many of you hate math but are forced to complete this course for graduation. Thus, I have attempted to maintain interest with "entertaining" word problems. They are not meant to be insulting to your intelligence. Obviously, most of the situations will never happen; however, calculus has several practical uses and such examples are found throughout Marsden and Weinstein's text. I would appreciate your comments on whether my "unusual" word problems should be kept for the next edition.

2.   (30,84)

3.   A vector has magnitude and direction; a scalar has magnitude only.

4.   (a) (-1,0)

(b) (-1,1)

(c) Yes; both are (-1,1) .

13.2 Vectors in Space

PREREQUISITES

1.   Recall how to do algebra with ordered pairs (Section 13.1).

2.   Recall how to do algebra with vectors in the plane (Section 13.1).

PREREQUISITE QUIZ

1.   Simplify the following expressions:

   (a)   (3,5) + (-1,2)

   (b)   -(1/2)(2,4)

2.   Let   P = (1,1) , Q = (-1,0) , and R = (2,-1) . What is $3\overrightarrow{PQ} - \overrightarrow{QR}$?

GOALS

1.   Be able to extend the ideas of the last section from ordered pairs to
   ordered triples.

2.   Be able to apply the concept of vectors to problem solving.

STUDY HINTS

1.   Right-hand rule. This will be important in Section 13.5 and also in
   physics courses. To determine the orientation, let the thumb of your
   right hand point in the positive z-direction. If your fingers, starting
   at the knuckles, curl from the positive x-axis to the positive y-axis,
   the orientation is right-handed. If not, it is left-handed. The right-
   handed orientation is the standard orientation.

2.   Ordered triples. The algebra is analogous to that of ordered pairs. The
   notation used to denote the vectors or ordered triples is the same as for
   ordered pairs.

3.   Notation. $\underline{i}$, $\underline{j}$, and $\underline{k}$ are the vectors with components $(1,0,0)$, $(0,1,0)$, and $(0,0,1)$. They point along the x , y , and z-axes, respectively. These are called the standard basis vectors. Similarly, $\underline{i}$ and $\underline{j}$ are $(1,0)$ and $(0,1)$ in the plane. Another special vector is $\underline{0}$, which is $(0,0,0)$, the zero vector.

4.   Applications. Up until now, we have been very loose with the usage of the terms speed and velocity. From now on, you should know that speed is a scalar and velocity is a vector. Also, distance is a scalar and displacement is a vector.

5.   Problem solving. Since vectors have magnitude and direction, they can be represented pictorially. Thus, you should normally sketch a diagram to help you visualize a vector word problem.

SOLUTIONS TO EVERY OTHER ODD EXERCISE

1.   The point $(1,0,0)$ is located on the x-axis.

5.   The sum of the ordered triples (x_1,y_1,z_1) and (x_2,y_2,z_2) is $(x_1 + x_2, y_1 + y_2, z_1 + z_2)$. Thus, $(6,0,5) + (5,0,6) = (11,0,11)$.

9.   If $\underline{v} = (1,-1,-1)$, then $2\underline{v} = (2,-2,-2)$ and $-\underline{v} = (-1,1,1)$.

13. The vector with components (a,b,c) can be written as $a\underline{i} + b\underline{j} + c\underline{k}$.
     Thus, the given vector is $-\underline{i} + 2\underline{j} + 3\underline{k}$.

17. The vector from $(0,1,2)$ to $(1,1,1)$ is $(1,1,1) - (0,1,2) =$
     $(1,0,-1)$, i.e., $\underline{i} - \underline{k}$.

21. 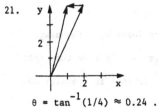   The vector joining the ship to the rock is
     $(2,4) - (1,0) = (1,4)$, i.e., $\underline{i} + 4\underline{j}$. As
     shown in the figure, $\tan \theta = 1/4$. Thus, the
     bearing of the rock from the ship is
     $\theta = \tan^{-1}(1/4) \approx 0.24$.

25. 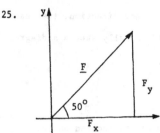   Using trigonometry, the horizontal com-
     ponent is $F_x = 50 \cos (50°) \approx 32.1$ lb ,
     and the vertical component is $F_y =$
     $50 \sin (50°) \approx 38.3$ lb .

29. 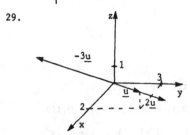   All three vectors lie on the same
     line passing through the origin.

33. The components of $\underline{v}$ are $(3,4,5)$ and the components of $\underline{w}$ are
     $(1,-1,1)$. Using scalar multiplication and addition of ordered triples,
     we get $6\underline{v} + 8\underline{w} = 6(3,4,5) + 8(1,-1,1) = (18,24,30) + (8,-8,8) =$
     $(26,16,38)$, which corresponds to the vector $26\underline{i} + 16\underline{j} + 38\underline{k}$.

37. $a\underline{v} + b\underline{w} = a(\underline{i} + \underline{j}) + b(-\underline{i} + \underline{j}) = (a - b)\underline{i} + (a + b)\underline{j}$. Therefore, we
     need to simultaneously solve $a - b = 3$ and $a + b = 7$. The solution
     is $a = 5$ and $b = 2$.

41. (a) Let x be the number of atoms of C , y be the number of atoms
          of H , and z be the number of atoms of O . Mathematically,
          the chemical equation is $p(3,4,3) + q(0,0,2) = r(1,0,2) + s(0,2,1)$.
     (b) We need to solve $3p = r$, $4p = 2s$, and $3p + 2q = 2r + s$.
          Suppose $p = 1$, then $r = 3$, $s = 2$, and $q = 5/2$. The
          smallest integers are obtained by multiplying by 2 , i.e., $p = 2$, $r = 6$, $s = 4$, and $q = 5$.
     (c)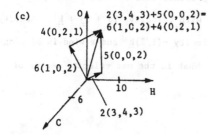

SECTION QUIZ

1. Which of the following notations can be used to represent a vector in
    space?
    (a) $-\underline{v}$
    (b) $\overrightarrow{RL}$
    (c) $\underline{n} + \underline{o}$
    (d) $(2,0,1)$
    (e) $(\underline{x},\underline{y},\underline{z})$

2. If a vector $\underline{v}$ originates at $(1,-2,1)$ and extends to $(0,3,2)$,
    write $-2\underline{v}$ with the standard basis notation.

3.   (a) Is north a vector? Why or why not?

    (b) Is velocity a vector? Why or why not?

4.   A motorboat in the middle of a bay is travelling in a northwesterly direction at a speed of 4 knots. A wind is blowing due south at 2 knots, and the current is flowing due east at 1 knot. What would be the motorboat's speed and direction in the absence of both the wind and the current?

5.   An old prospector had lost his compass and was lost in the middle of the desert when he came upon a rattlesnake. Keeping an eye on the rattler, the old prospector slowly backed into a cactus. The pain caused him to shoot into the air with velocity $2\underline{i} + 3\underline{j} + \underline{k}$. Gravity is causing him to descend with velocity $-(1/2)\underline{k}$ and a wind is blowing him away with velocity $2\underline{i} - \underline{j}$. What is the net velocity vector of the poor prospector?

ANSWERS TO PREREQUISITE QUIZ

1.   (a) (2,7)

    (b) (-1,-2)

2.   (-9,-2)

ANSWERS TO SECTION QUIZ

1.   a,b,c,d

2. $-2\underline{i} + 10\underline{j} - 2\underline{k}$

3.   (a) No; it is only a direction; it has no magnitude.

    (b) Yes; it has both magnitude and direction.

4.   Speed = $[21 + 12\sqrt{2}]^{1/2}$ knot ; direction = 0.66 radians north of west.

5. $4\underline{i} + 2\underline{j} + (1/2)\underline{k}$

13.3 Lines and Distance

PREREQUISITES

1. Recall how to perform vector algebra in space (Section 13.2).

2. Recall the triangle inequality (Section R.2).

3. Recall how to find the equations of lines and the distance between two points in the plane (Section R.4).

PREREQUISITE QUIZ

1. Let $\underline{u} = \underline{i} - \underline{j}$, $\underline{v} = 2\underline{j} + \underline{k}$, and $\underline{w} = \underline{i} + \underline{j} + \underline{k}$. What is $3\underline{u} - \underline{v} + 2\underline{w}$?

2. Let $(-1,-1)$, $(1,1)$, and $(0,1)$ be the vertices of a triangle.

   (a) Find the length of the three sides.

   (b) Show how the triangle inequality applies to this triangle.

3. What is the equation of the line passing through the points $(2,2)$ and $(3,1)$?

GOALS

1. Be able to write the equation of a line in the plane or in space given two points or a point and a direction.

2. Be able to normalize a given vector.

STUDY HINTS

1. Point-point form. You should memorize the formula for a line: $R = (1 - t)P + tQ$, where $R = (x,y,z)$, $P = (x_1,y_1,z_1)$, $Q = (x_2,y_2,z_2)$. To see if the direction is correct, plug in $t = 0$ — you should get the first point. Substituting $t = 1$ should give the second point. Look at Example 4 to see how the two point-point forms are equivalent.

2. **Point-direction form.** You should memorize $R = P + t\underline{d}$, where R , P , and $\underline{d}$ are defined according to the box on p. 663. From this, it is simple to get the component form. If you must find the direction vector $\underline{d}$ from two given points, substituting $t = 0$ and $t = 1$ to recover the given points will tell you if your direction $\underline{d}$ is correct.

3. **Lines in the plane.** The same equations hold if the lines lie in the xy-plane. Simply let $z = 0$.

4. **Non-intersecting lines in space.** Unlike lines in the plane, two lines in space will not generally intersect. See Example 6(a). As a side note, two non-intersecting lines in space are called **skew**.

5. **Length of a vector.** It is denoted by $\|\underline{v}\|$ and it is equal to $\sqrt{a^2 + b^2 + c^2}$ if $\underline{v} = a\underline{i} + b\underline{j} + c\underline{k}$. Note how this is similar to the length of the line segment drawn in the plane from $(0,0)$ to (a,b) , which is $\sqrt{a^2 + b^2}$.

6. **Properties of length.** Understand why the properties on p. 665 are reasonable to expect; with this understanding, you need not memorize them.

7. **Normalization.** If $\|\underline{u}\| = 1$, then $\underline{u}$ is called a unit vector. Dividing a vector $\underline{v}$ by its length $\|\underline{v}\|$ gives a unit vector, provided $\|\underline{v}\| \neq 0$. The process of creating a unit vector in this fashion is called normalization of $\underline{v}$.

8. **Distance formula.** As with the formula for length, we just insert an extra term to the planar formula; $+ (z_2 - z_1)^2$ in this case. The squaring process permits one to reverse the subscripts without changing the answer.

SOLUTIONS TO EVERY OTHER ODD EXERCISE

1. Let $\underline{v}_1$ be the vector joining A to B, and let $\underline{v}_2$ be the vector joining A to C. Then, the vector joining A to D is $(1/2)\underline{v}_1$ and the vector joining A to E is $(1/2)\underline{v}_2$. Thus, the vector joining E to D is $(1/2)\underline{v}_1 - (1/2)\underline{v}_2$, i.e., $(1/2)(\underline{v}_1 - \underline{v}_2)$. Since the vector joining C to B is $\underline{v}_1 - \underline{v}_2$, $\overline{DE}$ is parallel to $\overline{BC}$ and has one-half the length of $\overline{BC}$.

5. Let $\underline{v}_1$ be the vector from the origin to the point $(1,1)$ and $\underline{v}_2$ be the vector from the origin to the point $(2,-2)$. Then, from the results of Example 1, the vector from the origin to the midpoint of the line segment between $(1,1)$ and $(2,-2)$ is $(1/2)(\underline{v}_1 + \underline{v}_2) = (3/2,-1/2)$. Thus, the desired point is $(2/3)(3/2,-1/2) = (1,-1/3)$.

9. Using the method of Example 4, we obtain as the equation of the line $(x,y,z) = (1 - t)(0,0,0) + t(1,1,1) = (t,t,t)$. Alternatively, it can be expressed as $x = t$, $y = t$, and $z = t$.

13. This is analogous to Example 5(b). The line is described by $(x,y) = (-1,-2) + t(3,-2) = (-1 + 3t, -2 - 2t)$. Thus, $x = -1 + 3t$ and $y = -2 - 2t$.

17. If the lines intersect, then for some t_1 and t_2, we need $t_1 = 3t_2$, $3t_1 - 1 = 5$, and $4t_1 = 1 - t_2$. $3t_1 - 1 = 5$ implies $t_1 = 2$. Substituting $t_1 = 2$ into $t_1 = 3t_2$ yields $t_2 = 2/3$. However, substituting these values into the last equation gives $8 = 1/3$, which is false. Thus, the lines do not intersect.

21. The length of a vector is $\|\underline{v}\| = \|a\underline{i} + b\underline{j} + c\underline{k}\| = \sqrt{a^2 + b^2 + c^2}$. In this case, $\|\underline{v}\| = \sqrt{1 + 0 + 1} = \sqrt{2}$.

25. The length of the vector is $\| a\underline{i} - 3\underline{j} + \underline{k} \| = \sqrt{a^2 + 9 + 1} = \sqrt{10 + a^2} =$
    16 . Squaring yields $10 + a^2 = 256$, i.e., $a^2 = 246$, i.e., $a =$
    $\pm\sqrt{246}$.

29.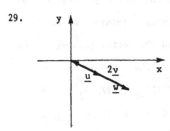
    $\underline{u} + 2\underline{v} + \underline{w} = \underline{0}$ implies $\underline{u} + \underline{w} = -2\underline{v}$.
    Since $\underline{u}$, $\underline{v}$, and $\underline{w}$ are unit vectors,
    we can only have $\underline{u} = \underline{w} = -\underline{v}$. A possible
    solution is $\underline{u} = \underline{i}$, $\underline{v} = -\underline{i}$, and
    $\underline{w} = \underline{i}$. Another possibility is shown in
    the diagram.

33. The normalization of $\underline{v}$ is $\underline{v}/\|\underline{v}\|$. For $\underline{v} = \underline{i} + \underline{j} + \underline{k}$, $\|\underline{v}\| =$
    $\sqrt{1 + 1 + 1} = \sqrt{3}$; the normalization is $(1/\sqrt{3})\underline{i} + (1/\sqrt{3})\underline{j} + (1/\sqrt{3})\underline{k}$.
    For $\underline{v} = \underline{i} + \underline{k}$, $\|\underline{v}\| = \sqrt{1 + 0 + 1} = \sqrt{2}$; the normalization is
    $(1/\sqrt{2})\underline{i} + (1/\sqrt{2})\underline{k}$.

37. The distance from (x_1, y_1, z_1) to (x_2, y_2, z_2) is $[(x_1 - x_2)^2 +$
    $(y_1 - y_2)^2 + (z_1 - z_2)^2]^{1/2}$. In this case, the distance is $[(1 - 1)^2 +$
    $(1 - 2)^2 + (2 - 3)^2]^{1/2} = \sqrt{2}$.

41. Let $\underline{v}_1$ be the velocity vector of the boat at full speed, i.e., $\underline{v}_1 =$
    $(0,12)$. Let $\underline{v}_2$ be the velocity vector of the current, i.e., $\underline{v}_2 =$
    $(5,0)$. Then the velocity vector of the boat in the current is $\underline{v}_1 +$
    $\underline{v}_2 = (5,12)$. Hence, the speed of the boat is $\|\underline{v}_1 + \underline{v}_2\| = \sqrt{5^2 + 12^2} =$
    13 knots.

45.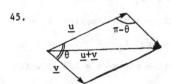
    Using the law of cosines, $\|\underline{u} + \underline{v}\|^2 =$
    $\|\underline{u}\|^2 + \|\underline{v}\|^2 - \|\underline{u}\|\|\underline{v}\| \cos(\pi - \theta) =$
    $(\|\underline{u}\| + \|\underline{v}\|)^2$ if $\cos(\pi - \theta) = -1$, i.e.,
    $\pi - \theta = \pi$ or $\theta = 0$. Equality holds if
    the angle between the two vectors is 0 , i.e., $\underline{u} = a\underline{v}$ for any scalar
    a . Using the vectors in Exercises 21 and 23, we have $\|\underline{u} + \underline{v}\| =$
    $\|3\underline{i} + 3\underline{k}\| = 3\sqrt{2}$ while $\|\underline{u}\| + \|\underline{v}\| = \sqrt{2} + 2\sqrt{2} = 3\sqrt{2} = \|\underline{u} + \underline{v}\|$. Equality
    is expected since $2\underline{i} + 2\underline{k} = 2(\underline{i} + \underline{k})$.

SECTION QUIZ

1. Write the equations describing the following lines:

   (a) The line from (1,3,2) to (-1,-3,-3).

   (b) The line containing (-1,2,1) with direction $2\underline{i} - 3\underline{j} + \underline{k}$.

2. True or false. The line (1,0,0) + t(0,1,0) does not contain (1,-1,0) because t would have to be negative.

3. For the line in Question 1(a), find a point on the line which is one unit distant from (1,3,2).

4. Find unit vectors which have the following directions:

   (a) $2\underline{i} + 3\underline{j} + \underline{k}$

   (b) $3\underline{j} - \underline{i} + \underline{k}$

   (c) $4\underline{j}$

5. Bert the Beaver had unusually fast-growing teeth, so he had to gnaw more wood than most other beavers. This was ideal for Bert because he lived on an island in the middle of the river. Since Bert gnawed more wood, it was easy for him to build a bridge to his island, which is known as Bert's Hill. The latest tree which fell due to Bert's gnawing originated at (2,3,1) and ended up at (3,1,2).

   (a) If the bridge follows a straight line, what is its equation?

   (b) What unit vector describes the bridge's direction?

   (c) How long is the bridge?

ANSWERS TO PREREQUISITE QUIZ

1. $5\underline{i} - 3\underline{j} + \underline{k}$

2. (a) $2\sqrt{2}$, 1, and $\sqrt{5}$

   (b) $1 + \sqrt{5} \geqslant 2\sqrt{2}$; $2\sqrt{2} + 1 \geqslant \sqrt{5}$; or $2\sqrt{2} + \sqrt{5} \geqslant 1$

3. $y = -x + 4$

ANSWERS TO SECTION QUIZ

1. (a) $(1,3,2) + (-2,-6,-5)t$

   (b) $(-1,2,1) + (2,-3,1)t$

2. False; t may have any real value.

3. $(1,3,2) \pm (2,6,5)/\sqrt{65}$

4. (a) $(2\underline{i} + 3\underline{j} + \underline{k})/\sqrt{14}$

   (b) $(-\underline{i} + 3\underline{j} + \underline{k})/\sqrt{11}$

   (c) $\underline{j}$

5. (a) $(2,3,1) + (1,-2,1)t$

   (b) $(\underline{i} - 2\underline{j} + \underline{k})/\sqrt{6}$

   (c) $\sqrt{6}$

13.4 <u>The Dot Product</u>

PREREQUISITES

1. Recall the algebraic and geometric properties of vectors (Sections 13.1 and 13.2).

2. Recall how to use the law of cosines (Section 5.1).

3. Recall how to use the inverse cosine function (Section 5.4).

PREREQUISITE QUIZ

1. What is the length of $\underline{v} = 2\underline{i} + \underline{j} - 2\underline{k}$?

2. 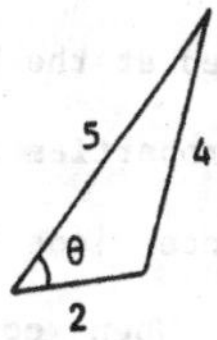   (a) Write the equation which describes the law of cosines.

             (b) Use the law of cosines to determine θ .

3. Compute $\cos^{-1}(\sqrt{3}/2)$.

4. Compute $\cos^{-1}(-1/2)$.

GOALS

1. Be able to state the definition of the dot product and compute it.

2. Be able to compute an orthogonal projection.

3. Be able to write the equation of a plane, given the appropriate data.

4. Be able to find the distance from a point to a line or a plane.

STUDY HINTS

1. <u>Dot product</u>. You should memorize the formulas $\underline{v}_1 \cdot \underline{v}_2 = a_1 a_2 + b_1 b_2 + c_1 c_2 = \|\underline{v}_1\| \cdot \|\underline{v}_2\| \cos \theta$ where $\underline{v}_1 = a_1\underline{i} + b_1\underline{j} + c_1\underline{k}$, $\underline{v}_2 = a_2\underline{i} + b_2\underline{j} + c_2\underline{k}$, and θ is the angle between $\underline{v}_1$ and $\underline{v}_2$. The dot product is also known as the scalar product or the inner product. Note that the

1. (continued)

    dot product is always a real number. To get the dot product in the
    plane, simply let $c_1 = c_2 = 0$.

2. <u>Geometry of dot product</u>. Since $\underline{v}_1 \cdot \underline{v}_2 = \|\underline{v}_1\| \cdot \|\underline{v}_2\| \cos \theta$, it is now
    possible to determine the angle θ between two vectors. If $\theta = 0$,
    we get the statement that $\|\underline{v}\|^2 = \underline{v} \cdot \underline{v}$. If $\theta = \pi/2$, $\cos \theta = 0$;
    therefore, two vectors are perpendicular when their dot product is 0 .
    The important formulas derived in this section are based upon this last
    fact.

3. <u>Dot product properties</u>. Properties 1 and 2, listed at the bottom of
    the box on p. 669, are obvious. The next three properties are analogous
    to properties of multiplication from algebra. Since $|\cos \theta| \leq 1$, we
    get property 6 from the geometric interpretation. When $\cos \theta = 0$,
    θ is $\pi/2$, so the vectors are perpendicular.

4. <u>Orthogonal projections</u>. The orthogonal projection of $\underline{v}$ on $\underline{u}$ is de-
    picted by placing $\underline{u}$ and $\underline{v}$ such that they originate from the same
    point. Then draw a perpendicular from the line defined by $\underline{u}$ and ex-
    tend it to the tip of $\underline{v}$. See Fig. 13.4.3. The projection is simply
    the "shadow" of $\underline{v}$ along $\underline{u}$; it is always a multiple of $\underline{u}$. It is
    probably best to remember the formula for the orthogonal projection of
    $\underline{v}$ on $\underline{u}$: $[(\underline{v} \cdot \underline{u})/(\underline{u} \cdot \underline{u})]\underline{u}$. As usual, understanding the formula will
    help you to remember it .

5. <u>Point to line distance</u>. Refer to Fig. 13.4.4. We know the coordinates
    of P and Q , so it is simple to determine $\|\overrightarrow{PQ}\|$. $\overrightarrow{PR}$ is simply an
    orthogonal projection, so $\|\overrightarrow{PR}\|$ can be determined. Finally, using the
    Pythagorean theorem, one can determine $\|\overrightarrow{RQ}\|$. You should get formula
    (4).

6. **Equation of a plane.** Let (x,y,z) and (x_0,y_0,z_0) be two points in the plane, so the vector from one point to the other is $\underline{v} = (x - x_0)\underline{i} + (y - y_0)\underline{j} + (z - z_0)\underline{k}$. If $\underline{n} = A\underline{i} + B\underline{j} + C\underline{k}$ is perpendicular to the plane, it is also perpendicular to $\underline{v}$, so $\underline{n} \cdot \underline{v} = 0$, which gives the equation of the plane, formula (5). Formula (6) simply substitutes D for $-Ax_0 - By_0 - Cz_0$. It is best to memorize these formulas.

7. **Plane through 3 points.** Three points not lying on a line determine a plane. To obtain the equation of the plane, remember that all three points satisfy $Ax + By + Cz + D = 0$. Solve the resulting three equations for A , B , and C in terms of D . Then choose D arbitrarily. (See Example 8.)

8. **Finding a plane's equation.** In general, three equations are needed to determine the numbers A , B , and C in the equation of the plane.

9. **Point to plane distance.** Equation (7) is derived in a fashion analogous to the derivation of Equation (4). $\underline{n}$ is chosen to be a unit vector in order to simplify things. This made $\underline{n} \cdot \underline{n} = 1$ and when we computed the length of the projection, we used $\|\underline{n}\| = 1$. Formula (7) is something you may wish to memorize rather than derive.

SOLUTIONS TO EVERY OTHER ODD EXERCISE

1. The dot product of $a_1\underline{i} + b_1\underline{j} + c_1\underline{k}$ and $a_2\underline{i} + b_2\underline{j} + c_2\underline{k}$ is $a_1a_2 + b_1b_2 + c_1c_2$. In this case, $(\underline{i} + \underline{j} + \underline{k}) \cdot (\underline{i} + \underline{j} + 2\underline{k}) = 1 \cdot 1 + 1 \cdot 1 + 1 \cdot 2 = 4$.

5. Use the formula $\theta = \cos^{-1}(\underline{v}_1 \cdot \underline{v}_2 / \|\underline{v}_1\| \|\underline{v}_2\|)$. Let $\underline{v}_1 = \underline{i} + \underline{j} + \underline{k}$, so $\|\underline{v}_1\| = \sqrt{3}$. Let $\underline{v}_2 = \underline{i} + \underline{j} + 2\underline{k}$, so $\|\underline{v}_2\| = \sqrt{6}$. Thus, $\theta = \cos^{-1}(4/\sqrt{3}\sqrt{6}) = \cos^{-1}(4/3\sqrt{2}) \approx 0.34$ radian.

9. We want $\underline{v} = a\underline{i} + b\underline{j}$ such that $2a - b = 0$ and $\sqrt{a^2 + b^2} = 1$. A
    simple method is to find an orthogonal vector and then normalize it.
    Such a vector is $\underline{i} + 2\underline{j}$, so $\underline{v} = (1/\sqrt{5})\underline{i} + (2/\sqrt{5})\underline{j}$.

13. From Example 4, the orthogonal projection of $\underline{v}$ on $\underline{u}$ is the vector
    $(\underline{v}\cdot\underline{u}/\underline{u}\cdot\underline{u})\underline{u}$. Thus, the length of this vector is $\| (\underline{v}\cdot\underline{u}/\underline{u}\cdot\underline{u})\underline{u}\| =$
    $|\underline{v}\cdot\underline{u}/\underline{u}\cdot\underline{u}|\|\underline{u}\|$. Since $\underline{u}\cdot\underline{u} = \|\underline{u}\|^2$, the length is $|\underline{v}\cdot\underline{u}|/\|\underline{u}\| =$
    $(|\underline{v}\cdot\underline{u}|/\|\underline{u}\|\|\underline{v}\|)\|\underline{v}\| = |\cos\theta|\|\underline{v}\|$, where θ is the angle between $\underline{u}$
    and $\underline{v}$.

17. Use the formula $\text{dist}(Q,\ell) = \{(x_1 - x_0)^2 + (y_1 - y_0)^2 + (z_1 - z_0)^2 -$
    $[a(x_1 - x_0) + b(y_1 - y_0) + c(z_1 - z_0)]^2/(a^2 + b^2 + c^2)\}^{1/2}$. The line
    $x = 3t + 2$, $y = -t - 1$, $z = t + 1$ goes through $(2,-1,1) =$
    (x_0,y_0,z_0) and has direction $(3,-1,1) = (a,b,c)$. With $(x_1,y_1,z_1) =$
    $(1,1,2)$, the distance is $\{(-1)^2 + (-2)^2 + (1)^2 - [3(-1) + (-1)(-2) +$
    $(1)(1)]^2/(9 + 1 + 1)\}^{1/2} = \sqrt{6 - 0/11} = \sqrt{6}$.

21. The equation of the plane through (x_0,y_0,z_0) with normal vector
    $A\underline{i} + B\underline{j} + C\underline{z}$ is $A(x - x_0) + B(y - y_0) + C(z - z_0) = 0$. Thus, the
    plane is $1(x - 0) + 0(y - 0) + 0(z - 0) = 0$, i.e., $x = 0$.

25. From the coefficients, a normal vector is $2\underline{i} + 3\underline{j} + \underline{k}$. Its length is
    $\sqrt{4 + 9 + 1} = \sqrt{14}$, so a normal unit vector is $(2/\sqrt{14})\underline{i} + (3/\sqrt{14})\underline{j} +$
    $(1/\sqrt{14})\underline{k}$.

29. Since the general equation of the plane is $Ax + By + Cz + D = 0$, we
    can substitute the given points to get $B + D = 0$, $A + D = 0$, and
    $C + D = 0$. Therefore $D = -A = -B = -C$. Arbitrarily choosing $D =$
    -1 yields $A = B = C = 1$, and the equation of the plane becomes
    $x + y + z - 1 = 0$.

33. The distance from (x_1, y_1, z_1) to the plane $Ax + By + Cz + D = 0$ is $|Ax_1 + By_1 + Cz_1 + D|/\sqrt{A^2 + B^2 + C^2}$. The line is parametrically represented by $x = 2t$, $y = -t$, and $z = 3t$. It meets the plane when $2x - y + 3z = 4t + t + 9t = 7$, i.e., $14t = 7$, i.e., $t = 1/2$. The point of intersection is $(1, -1/2, 3/2)$. The distance from $(0,0,0)$ to the plane is $|(2)(0) - (1)(0) + (3)(0) - 7|/\sqrt{4 + 1 + 9} = 7/\sqrt{14} = \sqrt{14}/2$.

37. Since the distance from (x_1, y_1, z_1) to the plane $Ax + By + Cz + D = 0$ is $|Ax_1 + By_1 + Cz_1 + D|/\sqrt{A^2 + B^2 + C^2}$, we have, in this particular case, distance $= |(1)(1) - (1)(1) - (1)(1) + 10|/\sqrt{1 + 1 + 1} = 9/\sqrt{3} = 3\sqrt{3}$.

41.

Let the coordinates of the two points be (a,b) and (c,d) . The line that we are looking for contains the point $((a + c)/2,$ $(b + d)/2)$ and has normal vector $(a - c)\underline{i} + (b - d)\underline{j}$. Thus, the equation of the line has the form $(a - c)x + (b - d)y = E$. Since the midpoint $((a + c)/2, (b + d)/2)$ is on the line, we have $(a - c)((a + c)/2) + (b - d)((b + d)/2) = E$. So the equation of the line is $(a - c)x + (b - d)y = (1/2)(a^2 - c^2 + b^2 - d^2)$. We need to show that any (x,y) satisfying the above equation is equidistant from the two points (a,b) and (c,d) . We have $y = -[(a - c)/(b - d)]x + [(a^2 - c^2 + b^2 - d^2)/2(b - d)]$, so $(x - a)^2 + (y - b)^2 = (x - a)^2 + \{-[(a - c)/(b - d)]x + (a^2 - c^2 + b^2 - d^2)/2(b - d) - b\}^2 = (x - a)^2 + [((a^2 - 2ax + x^2) - (c^2 - 2cx + x^2) - (b^2 - 2bd + d^2))/2(b - d)]^2 = (x - a)^2 + [((a - x)^2 - (c - x)^2 - (b - d)^2)/2(b - d)]^2$. On the other hand, $(x - c)^2 + (y - d)^2 = (x - c)^2 + [((a - x)^2 - (c - x)^2 + (b - d)^2)/2(b - d)]^2$. Putting the

41.   (continued)

last two expressions over a common denominator $4(b - d)^2$, we see that

they are both equal to $[(a - x)^4 + (c - x)^4 + (b - d)^4 - 2(a - x)^2(c - x)^2 +$

$2(a - x)^2(b - d)^2 + 2(c - x)^2(b - d)^2]/4(b - d)^2$. Thus, (x,y) on the

line is indeed equidistant from (a,b) and (c,d) .

45.   Let $P_1 = (p,q)$ and $P_2 = (r,s)$. We need $(x - p)^2 + (y - q)^2 =$

$(x - r)^2 + (y - s)^2$,   so $(x - p)^2 - (x - r)^2 = (y - s)^2 - (y - q)^2$.

Factoring gives $(x - p + x - r)(x - p - x + r) = (y - s + y - q) \times$

$(y - s - y + q)$,   so $(2x - p - r)(r - p) = (2y - s - q)(q - s)$,   so

$(2x)(r - p) + p^2 - r^2 = 2y(q - s) + s^2 - q^2$,   so $2x(r - p) +$

$2y(s - q) = r^2 + s^2 - p^2 - q^2$. Hence, $ax + by = c$ if $a = r - p$,

$b = s - q$,   and $c = (r^2 + s^2 - p^2 - q^2)/2$.

49.   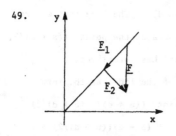 A unit vector in the direction of the plane

is $\underline{e}_1 = (1/\sqrt{2})(-\underline{i} - \underline{j})$,   and one perpen-

dicular to it is $\underline{e}_2 = (1/\sqrt{2})(\underline{i} - \underline{j})$.

Write $\underline{F} = -F\underline{j}$,   where $F = \|\underline{F}\|$ is the

magnitude of $\underline{F}$. Then we write $\underline{F} =$

$\alpha\underline{e}_1 + \beta\underline{e}_2 = (\alpha/\sqrt{2})(-\underline{i} - \underline{j}) + (\beta/\sqrt{2})(\underline{i} - \underline{j}) =$

$-F\underline{j}$.   This will hold if $-\alpha + \beta = 0$ and $-\alpha/\sqrt{2} - \beta/\sqrt{2} = -1$,   i.e.,

if $\alpha = \beta$ and $\alpha + \beta = \sqrt{2}$,   i.e., $\alpha = \beta = 1/\sqrt{2}$. Thus, $\underline{F} = \underline{F}_1 + \underline{F}_2$,

where $\underline{F}_1 = -(F/2)(\underline{i} + \underline{j})$ and $\underline{F}_2 = (F/2)(\underline{i} - \underline{j})$.

53.   Let $\underline{u} = x\underline{i} + y\underline{j} + z\underline{k}$.

(a) $\underline{u} \cdot \underline{u} = x^2 + y^2 + z^2 \geqslant 0$.

(b)   If $\underline{u} \cdot \underline{u} = 0$, then $x^2 + y^2 + z^2 = 0$,   so $x = y = z = 0$,   and

$\underline{u} = \underline{0}$.

(c)   Let $\underline{v} = r\underline{i} + s\underline{j} + t\underline{k}$. Then $\underline{u} \cdot \underline{v} = xr + ys + zt$ and $\underline{v} \cdot \underline{u} = rx +$

$sy + tz = \underline{u} \cdot \underline{v}$.

53. (d) Let $\underline{w} = \ell\underline{i} + m\underline{j} + n\underline{k}$. Then $(a\underline{u} + b\underline{v})\cdot\underline{w} = [(ax + br)\underline{i} +$

$(ay + bs)\underline{j} + (az + bt)\underline{k}]\cdot(\ell\underline{i} + m\underline{j} + n\underline{k}) = (a\ell x + b\ell r)\underline{i} +$

$(amy + bms)\underline{j} + (anz + bnt)\underline{k}$. Also $a(\underline{u}\cdot\underline{w}) + b(\underline{v}\cdot\underline{w}) =$

$a(x\ell + ym + zn) + b(r\ell + sm + tn) = (a\ell x + b\ell r)\underline{i} + (amy + bms)\underline{j} +$

$(anz + bnt)\underline{k} = (a\underline{u} + b\underline{v})\cdot\underline{w}$.

57. (a) Equating components, we get $\mu = a/\sqrt{a^2 + b^2 + c^2}$, $\lambda = b/$

$\sqrt{a^2 + b^2 + c^2}$, and $\nu = c/\sqrt{a^2 + b^2 + c^2}$. Hence, if $s =$

$t\sqrt{a^2 + b^2 + c^2}$, then $P_0 + s(\mu,\lambda,\nu) = P_0 + t\sqrt{a^2 + b^2 + c^2} \times$

$(a/\sqrt{a^2 + b^2 + c^2}, b/\sqrt{a^2 + b^2 + c^2}, c/\sqrt{a^2 + b^2 + c^2}) = P_0 +$

$t(a,b,c)$, which is the same line.

(b) Note that $\|\underline{u}\| = \|\underline{d}\|/\|\underline{d}\| = 1$. Hence, $\underline{i}\cdot\underline{u} = \mu = \|\underline{i}\|\|\underline{u}\| \cos \alpha =$

$\cos \alpha$; $\underline{j}\cdot\underline{u} = \lambda = \|\underline{j}\|\|\underline{u}\| \cos \beta = \cos \beta$; and $\underline{k}\cdot\underline{u} = \nu =$

$\|\underline{k}\|\|\underline{u}\| \cos \gamma = \cos \gamma$.

(c) $\cos^2\alpha + \cos^2\beta + \cos^2\gamma = \mu^2 + \lambda^2 + \nu^2 = \underline{u}\cdot\underline{u} = \|\underline{u}\|^2 = 1$.

(d) For L_1 , $\cos \alpha = 1/\sqrt{3}$ so $\alpha = \cos^{-1}(1/\sqrt{3})$; $\cos \beta = 1/\sqrt{3}$, so

$\beta = \cos^{-1}(1/\sqrt{3})$; and $\cos \gamma = 1/\sqrt{3}$, so $\gamma = \cos^{-1}(1/\sqrt{3})$. L_2

has the same direction angles and cosines as L_1 , since its

direction is the same. For L_3 , $\cos \alpha = 1/\sqrt{83}$, so $\alpha =$

$\cos^{-1}(1/\sqrt{83})$; $\cos \beta = 1/\sqrt{83}$, so $\beta = \cos^{-1}(1/\sqrt{83})$; and

$\cos \gamma = 9/\sqrt{83}$, so $\gamma = \cos^{-1}(9/\sqrt{83})$. L_4 has the same direction

angles and cosines as L_3 , since it has the same direction.

(e) $\alpha = \beta = \gamma$ implies $\cos \alpha = \cos \beta = \cos \gamma$. Thus, $\mu = \lambda = \nu$,

so $a = b = c$. Hence only the line $t(1,1,1)$ has direction

angles $\alpha = \beta = \gamma$.

SECTION QUIZ

1.   Compute the following dot products:

   (a) $(\underline{i} + \underline{j} - 3\underline{k}) \cdot (-2\underline{j} - \underline{k})$

   (b) $(2\underline{i} + 3\underline{j}) \cdot (\underline{i})$

2.   Let ℓ be the line $(1,5,-2) + t(-2,3,1)$. What is an equation of a
     line perpendicular to ℓ and which passes through $(5,-1,6)$?

3.   A plane P contains the points $(1,1,1)$, $(2,3,0)$, and $(2,5,-1)$.
     Find the equation of a parallel plane containing the origin.

4.   In the plane, find y if the vector $2\underline{i} + y\underline{j}$ is $\pi/3$ radians from the
     vector $-3\underline{i} + 4\underline{j}$.

5.   Let $\underline{u} = (1,2,3)$ and $\underline{v} = (4,3,2)$. What is the orthogonal projection
     of $\underline{v}$ on $\underline{u}$ and the orthogonal projection of $\underline{u}$ on $\underline{v}$?

6.   A proud unicorn has its horn extending from the point $P(2,1,1)$ to the
     point $Q(3,2,3)$. A magic mirror containing the image of an evil wizard
     exists on the plane $2x - 3y + z = 0$.

   (a)   Suppose the evil wizard's ugly face provokes the unicorn into
         attacking. How far does the unicorn need to move, i.e., what is
         the distance from Q to the plane?

   (b)   The sun is situated so that a shadow is cast on the ground in the
         direction $-\underline{i} - \underline{j}$. How long is the shadow, i.e., what is the
         length of the projection of the horn's vector upon $-\underline{i} - \underline{j}$?

ANSWERS TO PREREQUISITE QUIZ

1.   3

2.   (a) $a^2 + b^2 - 2ab \cos \theta = c^2$

     (b) $\theta = \cos^{-1}(13/20) \approx 0.86$

3. $\pi/6$

4. $2\pi/3$

ANSWERS TO SECTION QUIZ

1.   (a) 1

      (b) 2

2. $(5,-1,6) + (6,-9,39)t$ is one possible answer.

3. $y + 2z = 0$

4. $y = (96 \pm 50\sqrt{3})/39$

5. $(16/\sqrt{14})(1,2,3)$ and $(16/\sqrt{29})(4,3,2)$

6.   (a) $3/\sqrt{14}$

      (b) $\sqrt{2}$

13.5 The Cross Product

PREREQUISITES

1.   Recall how to compute the equation of a plane, given three points in the plane (Section 13.4).

2.   Recall the right-hand rule (Section 13.2).

PREREQUISITE QUIZ

1.   Find the equation of the plane passing through the points (1,0,0) , (0,3,0) , and (-1,2,3) .

2. 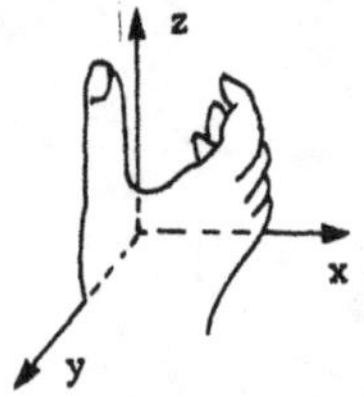   In the diagram, the right hand is the shown. Do the x-, y-, and z-axes form a right-handed or a left-handed system?

GOALS

1.   Be able to state the definition of the cross product and compute it.

2.   Be able to use the cross product to find the area of a parallelogram and to find the equation of a plane.

STUDY HINTS

1.   Cross products. The cross product, also known as the vector product, is always a vector, unlike the dot product, which is a scalar. Do not memorize the component formula. You will be given an easier formula in the next section. For now, keep referring to formula (1) or use the distributive property of cross products along with Fig. 13.5.5. Note the clockwise positioning of $\underline{i}$, $\underline{j}$, and $\underline{k}$ in the figure.

2. <u>Cross product geometry</u>. The cross product $\underline{u} \times \underline{v}$ is perpendicular to both $\underline{u}$ and $\underline{v}$. $\underline{u}$, $\underline{v}$, and $\underline{u} \times \underline{v}$, in this order, form a right-handed system. Also, the length of the cross product is the area of the parallelogram spanned by $\underline{u}$ and $\underline{v}$. Note that the cross product is related to $\sin \theta$, whereas the dot product is related to $\cos \theta$, where θ is the angle between the two vectors.

3. <u>Algebraic properties</u>. If the cross product is zero, then either: (i) the length of one of the vectors must be zero, or (ii) $\sin \theta = 0$, i.e., $\theta = 0$, i.e., the vectors must be parallel. Example 2(a) clearly demonstrates that the cross product is <u>not</u> associative. Also, it is <u>not</u> commutative; changing the multiplication order changes the sign, i.e., $\underline{u} \times \underline{v} = -\underline{v} \times \underline{u}$. However, the distributive property does apply for cross products.

4. <u>The equation of a plane</u>. In Section 13.4, when you were given three points in a plane, you had to solve three simultaneous equations. The three points not lying on a line define two vectors in the plane. Their cross product is perpendicular to both vectors, and therefore, it is normal to the plane. If this normal is $\underline{n} = A\underline{i} + B\underline{j} + C\underline{k}$, the plane is $Ax + By + Cz + D = 0$. You get D by substituting one of the points.

SOLUTIONS TO EVERY OTHER ODD EXERCISE

1. Use the algebraic rules for dot products along with the following:
$\underline{i} \times \underline{i} = \underline{0}$, $\underline{i} \times \underline{j} = \underline{k}$, $\underline{i} \times \underline{k} = -\underline{j}$, $\underline{j} \times \underline{i} = -\underline{k}$, $\underline{j} \times \underline{j} = \underline{0}$, $\underline{j} \times \underline{k} = \underline{i}$, $\underline{k} \times \underline{i} = \underline{j}$, $\underline{k} \times \underline{j} = -\underline{i}$, and $\underline{k} \times \underline{k} = \underline{0}$. Thus, $(\underline{i} - \underline{j} + \underline{k}) \times (\underline{j} - \underline{k}) = \underline{i} \times \underline{j} - \underline{j} \times \underline{j} + \underline{k} \times \underline{j} - \underline{i} \times \underline{k} + \underline{j} \times \underline{k} - \underline{k} \times \underline{k} = \underline{k} - \underline{0} - \underline{i} + \underline{j} + \underline{i} - \underline{0} = \underline{j} + \underline{k}$.

5. Use the method of Exercise 1. $(3\underline{i} + 2\underline{j}) \times 3\underline{j} = 9(\underline{i} \times \underline{j}) + 6(\underline{j} \times \underline{j}) = 9\underline{k} + \underline{0}$. Thus, $[(3\underline{i} + 2\underline{j}) \times 3\underline{j}] \times (2\underline{i} - \underline{j} + \underline{k}) = 9\underline{k} \times (2\underline{i} - \underline{j} + \underline{k}) = 18(\underline{k} \times \underline{i}) - 9(\underline{k} \times \underline{j}) + 9(\underline{k} \times \underline{k}) = 9\underline{i} + 18\underline{j}$.

9. Use the method of Exercise 1 to get $(\underline{i} + \underline{k}) \times (\underline{i} + \underline{j} + \underline{k}) = \underline{i} \times \underline{i} + \underline{k} \times \underline{i} + \underline{i} \times \underline{j} + \underline{k} \times \underline{j} + \underline{i} \times \underline{k} + \underline{k} \times \underline{k} = \underline{0} + \underline{j} + \underline{k} - \underline{i} - \underline{j} + \underline{0} = -\underline{i} + \underline{k}$.

13. The area of the parallelogram spanned by $\underline{v}_1$ and $\underline{v}_2$ is $\|\underline{v}_1 \times \underline{v}_2\|$. In this case, $\underline{i} \times (\underline{i} - 2\underline{j}) = \underline{i} \times \underline{i} - 2\underline{i} \times \underline{j} = \underline{0} - 2\underline{k} = -2\underline{k}$. Thus, the area is $\|-2\underline{k}\| = 2$.

17. An orthogonal unit vector is $\underline{v}_1 \times \underline{v}_2 / \|\underline{v}_1 \times \underline{v}_2\|$. $(\underline{i} - \underline{j} - \underline{k}) \times (2\underline{i} - 2\underline{j} + \underline{k}) = 2\underline{i} \times \underline{i} - 2\underline{j} \times \underline{i} - 2\underline{k} \times \underline{i} - 2\underline{i} \times \underline{j} + 2\underline{j} \times \underline{j} + 2\underline{k} \times \underline{j} + \underline{i} \times \underline{k} - \underline{j} \times \underline{k} - \underline{k} \times \underline{k} = \underline{0} + 2\underline{k} - 2\underline{j} - 2\underline{k} + \underline{0} - 2\underline{i} - \underline{j} - \underline{i} - \underline{0} = -3\underline{i} - 3\underline{j}$ and $\|-3\underline{i} - 3\underline{j}\| = \sqrt{9 + 9 + 0} = 3\sqrt{2}$. Thus, the desired vector is $(-1/\sqrt{2})\underline{i} - (1/\sqrt{2})\underline{j}$.

21. $2\underline{i} - \underline{k}$ and $4\underline{j} - 3\underline{k}$ are vectors in the desired plane. $(2\underline{i} - \underline{k}) \times (4\underline{j} - 3\underline{k})$ is normal to the plane, and it equals $8\underline{i} \times \underline{j} - 4\underline{k} \times \underline{j} - 6\underline{i} \times \underline{k} + 3\underline{k} \times \underline{k} = 8\underline{k} + 4\underline{i} + 6\underline{j} + \underline{0}$. The plane has the form $4x + 6y + 8z + D = 0$. Substituting $(0,0,0)$ yields $D = 0$, so the plane is $4x + 6y + 8z = 0$, i.e., $2x + 3y + 4z = 0$.

25. The area of the triangle is half of the area of the parallelogram which has the same vertices. The vector from $(0,1,2)$ to $(3,4,5)$ is $\underline{v}_1 = 3\underline{i} + 3\underline{j} + 3\underline{k}$, and the vector from $(0,1,2)$ to $(-1,-1,0)$ is $\underline{v}_2 = -\underline{i} - 2\underline{j} - 2\underline{k}$. $\underline{v}_1 \times \underline{v}_2 = -3\underline{i} \times \underline{i} - 3\underline{j} \times \underline{i} - 3\underline{k} \times \underline{i} - 6\underline{i} \times \underline{j} - 6\underline{j} \times \underline{j} - 6\underline{k} \times \underline{j} - 6\underline{j} \times \underline{k} - 6\underline{k} \times \underline{k} = \underline{0} + 3\underline{k} - 3\underline{j} - 6\underline{k} - \underline{0} + 6\underline{i} + 6\underline{j} - 6\underline{i} - \underline{0} = 3\underline{j} - 3\underline{k}$. $\|\underline{v}_1 \times \underline{v}_2\| = \sqrt{0 + 9 + 9} = 3\sqrt{2}$, so the area of the triangle is $3\sqrt{2}/2$.

29. Let $\underline{v}_1 = b_1\underline{i} + c_1\underline{j} + d_1\underline{k}$ and $\underline{v}_2 = b_2\underline{i} + c_2\underline{j} + d_2\underline{k}$. By the component formula, $(a\underline{v}_1) \times \underline{v}_2 = (ac_1d_2 - ad_1c_2)\underline{i} + (ad_1b_2 - ab_1d_2)\underline{j} +$
$(ab_1c_2 - ac_1b_2)\underline{k}$. The right-hand side is $a(\underline{v}_1 \times \underline{v}_2) =$
$a[(c_1d_2 - d_1c_2)\underline{i} + (d_1b_2 - b_1d_2)\underline{j} + (b_1c_2 - c_1b_2)\underline{k}] = (ac_1d_2 - ad_1c_2)\underline{i} +$
$(ad_1b_2 - ab_1d_2)\underline{j} + (ab_1c_2 - ac_1b_2)\underline{k}$.

33. Apply the result of Exercise 32. The direction $\underline{d}$ of the line is
$-b\underline{i} + a\underline{j}$ and $(0,c/b)$ is a point P_0 on the line. Hence, $x\underline{i} +$
$(y - c/b)\underline{j}$ is a vector from $(0,c/b)$ to (x,y) . The distance is
$\|(x\underline{i} + (y - c/b)\underline{j}) \times (-b\underline{i} + a\underline{j})\|/\|-b\underline{i} + a\underline{j}\| = \|-bx\underline{i} \times \underline{i} -$
$b(y - c/b)\underline{j} \times \underline{i} + ax\underline{i} \times \underline{j} + a(y - c/b)\underline{j} \times \underline{j}\|/\sqrt{a^2 + b^2} = \|-(c - by)\underline{k} +$
$ax\underline{k}\|/\sqrt{a^2 + b^2} = |ax + by - c|/\sqrt{a^2 + b^2}$.

37. Since all vectors in this problem are unit vectors, $\|\underline{N} \times \underline{a}\| = \sin \theta_1$
and $\|\underline{N} \times \underline{b}\| = \sin \theta_2$. From Snell's law, $n_1 \sin \theta_1 = n_2 \sin \theta_2$.
Hence, $n_1\|\underline{N} \times \underline{a}\| = n_2\|\underline{N} \times \underline{b}\|$.

    In order to establish that $\underline{N} \times \underline{a}$ and $\underline{N} \times \underline{b}$ have the same direc-
tion, we assume that $\underline{N}$, $\underline{a}$, and $\underline{b}$ all lie in the same plane, and
$\underline{a}$ and $\underline{b}$ are on the same side of $\underline{N}$. Hence, $\underline{N} \times \underline{a}$ and $\underline{N} \times \underline{b}$
both are perpendicular to this plane and parallel to each other. Thus,
$n_1\|\underline{N} \times \underline{a}\|$ and $n_2\|\underline{N} \times \underline{b}\|$ have the same direction, as well as the same
magnitude, and so are equal.

41.

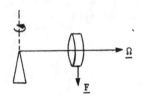

The direction in which the flywheel turns
is $\underline{\Omega} \times \underline{F}$. As shown in the diagram, the
gyroscope will rotate to the left (as viewed
from above).

SECTION QUIZ

1. A parallelogram has vertices at $(5,2,1)$, $(0,-1,1)$, and $(3,-1,4)$.
    What is the area of the parallelogram?

2. $\underline{u} \times \underline{v} = \underline{0}$ has what geometric interpretation? Assume that neither $\underline{u}$
    nor $\underline{v}$ are $\underline{0}$.

3. Use the cross product to find the plane containing $(3,1,1)$, $(2,3,0)$,
    and $(2,3,-1)$.

4. Compute the following cross products:

    (a) $(3\underline{i} + \underline{j}) \times \underline{k}$

    (b) $(2\underline{k}) \times (2\underline{i} - \underline{k})$

    (c) $(\underline{i} + \underline{j} + \underline{k}) \times (3\underline{i} + 2\underline{j} - 2\underline{k})$

5. A certain city, which was designated as the worst city in the country to
    live in, has been left unprotected by the Defense Department. Con-
    sequently, the city must provide its own defenses. Their missile neu-
    tralizer works only at a perpendicular to the base of the weapon. If,
    at a certain instant, two vectors along the base are $2\underline{i} + 3\underline{j}$ and
    $\underline{i} - \underline{j} + 2\underline{k}$, what is the line along which a nuclear warhead can be
    neutralized? The neutralizer originates at $(1,1,0)$.

ANSWERS TO PREREQUISITE QUIZ

1. $9x + 3y + 4z = 9$

2. Left-handed

ANSWERS TO SECTION QUIZ

1. $\sqrt{387}$

2. $\underline{u}$ and $\underline{v}$ are parallel.

3. $-2x - y + 7 = 0$

4.   (a) $\underline{i} - 3\underline{j}$

      (b) $4\underline{j}$

      (c) $-4\underline{i} + 5\underline{j} - \underline{k}$

5. $(1,1,0) + (6,-4,-5)t$

13.6 Matrices and Determinants

PREREQUISITES

1.   Recall how to compute a cross product (Section 13.5).

PREREQUISITE QUIZ

1.   Compute $\underline{i} \times \underline{j}$.

2.   Compute $(2\underline{i} - 3\underline{k}) \times (\underline{j} + \underline{k})$.

GOALS

1.   Be able to compute 2×2 and 3×3 determinants.

2.   Be able to use determinants for computing cross products.

STUDY HINTS

1.   Determinants. The determinant of a matrix is a number whose absolute
     value is the area of the parallelogram or the volume of the parallele-
     piped spanned by a given set of 2 or 3 vectors. Note that the com-
     ponents of the vector are listed across a row. Also note that deter-
     minants always have a square configuration.

2.   Orientation. A positive determinant tells you that two vectors which
     make up a determinant have a counterclockwise orientation or that the
     three vectors form a right-handed system.

3.   Matrices. A matrix is simply a rectangular (not necessarily square)
     array of numbers. It is not a number. The usefulness of matrices will
     be seen in Chapter 15. Two matrices are equal if and only if corres-
     ponding components are equal. Compare this with determinants which may
     be equal even though the components are not equal.

4. Notation. Determinants are represented by numbers within vertical bars.
    Matrices are numbers within brackets.

5. Computing 2 × 2 determinants. You should memorize the formula
    $\begin{vmatrix} a & b \\ c & d \end{vmatrix}$ = ad - bc . Just take the product of the diagonal going left
    to right and subtract the product of the diagonal going in the opposite
    direction.

6. Computing 3 × 3 determinants. Use the checkerboard pattern shown on
    p. 687 which begins with a plus sign in the upper left corner. Choose
    any column or row — usually picking the one with the most zeros saves
    work. Draw vertical and horizontal lines through the first number of
    the row or column. The numbers remaining form a 2 × 2 determinant,
    which should be multiplied by the number (with sign determined by the
    checkerboard) through which both lines were drawn. Repeat for the re-
    maining numbers of the row or column. Finally, sum the results. This
    process, called expansion by minors, works for any row or column . Be
    sure to use the correct sign.

7. Cross products and determinants. The cross product can be computed as
    a 3 × 3 determinant. This is the easiest way to remember how to com-
    pute the cross product. The determinant is written with $\underline{i}$, $\underline{j}$, and
    $\underline{k}$ across the top row. The components of the first vector is written
    across the middle row and the components of the second vector is on the
    bottom row. See Example 5(b).

8. Triple products. The volume of the parallelepiped spanned by $\underline{v}_1$,
    $\underline{v}_2$, and $\underline{v}_3$ may be computed using the triple product $(\underline{v}_1 \times \underline{v}_2) \cdot \underline{v}_3$.
    Example 8 demonstrates an interesting fact about triple products.

SOLUTIONS TO EVERY OTHER ODD EXERCISE

1. Using the fact that $\begin{vmatrix} a & b \\ c & d \end{vmatrix} = ad - bc$, we have $\begin{vmatrix} 1 & 1 \\ -1 & 1 \end{vmatrix} = (1)(1) -$ $(-1)(1) = 2$.

5. Since $\begin{vmatrix} a & b \\ c & d \end{vmatrix} = ad - bc$, we get $\begin{vmatrix} 1 & 2 \\ 3 & 4 \end{vmatrix} = (1)(4) - (2)(3) = -2$.

9. Since $\begin{vmatrix} a & b \\ c & d \end{vmatrix} = ad - bc$, we get $\begin{vmatrix} a & b \\ 0 & c \end{vmatrix} = ac = b(0) = ac$.

13. The left-hand side is $\begin{vmatrix} ra & rb \\ c & d \end{vmatrix} = rad - rbc$ and the right-hand side is $r \begin{vmatrix} a & b \\ c & d \end{vmatrix} = r(ad - bc) = rad - rbc$. Therefore, the identity is proven.

17. Expand in minors of the first row. $\begin{vmatrix} 2 & -1 & 0 \\ 4 & 3 & 2 \\ 3 & 0 & 1 \end{vmatrix} = 2 \begin{vmatrix} 3 & 2 \\ 0 & 1 \end{vmatrix} + 1 \begin{vmatrix} 4 & 2 \\ 3 & 1 \end{vmatrix} + 0 \begin{vmatrix} 4 & 3 \\ 3 & 0 \end{vmatrix} =$ $2(3) - 2 + 0 = 4$.

21. Expand in minors of the first row to get $\begin{vmatrix} 2 & 1 & 0 \\ 0 & 2 & 1 \\ 1 & 0 & 2 \end{vmatrix} = 2 \begin{vmatrix} 2 & 1 \\ 0 & 2 \end{vmatrix} - 1 \begin{vmatrix} 0 & 1 \\ 1 & 2 \end{vmatrix} +$ $0 \begin{vmatrix} 0 & 2 \\ 1 & 0 \end{vmatrix} = 2(4) - 1(-1) + 0 = 9$.

25. In determinant form, $(a_1 \underline{i} + b_1 \underline{j} + c_1 \underline{k}) \times (a_2 \underline{i} + b_2 \underline{j} + c_2 \underline{k})$ is $\begin{vmatrix} \underline{i} & \underline{j} & \underline{k} \\ a_1 & b_1 & c_1 \\ a_2 & b_2 & c_2 \end{vmatrix}$. Thus, $(3\underline{i} - \underline{j}) \times (\underline{j} + \underline{k}) = \begin{vmatrix} \underline{i} & \underline{j} & \underline{k} \\ 3 & -1 & 0 \\ 0 & 1 & 1 \end{vmatrix} = \begin{vmatrix} -1 & 0 \\ 1 & 1 \end{vmatrix} \underline{i} -$ $\begin{vmatrix} 3 & 0 \\ 0 & 1 \end{vmatrix} \underline{j} + \begin{vmatrix} 3 & -1 \\ 0 & 1 \end{vmatrix} \underline{k} = -\underline{i} - 3\underline{j} + 3\underline{k}$.

29. By the method of Exercise 25, we get $(\underline{i} - \underline{k}) \times (\underline{i} + \underline{k}) = \begin{vmatrix} \underline{i} & \underline{j} & \underline{k} \\ 1 & 0 & -1 \\ 1 & 0 & 1 \end{vmatrix} =$ $\begin{vmatrix} 0 & -1 \\ 0 & 1 \end{vmatrix} \underline{i} - \begin{vmatrix} 1 & -1 \\ 1 & 1 \end{vmatrix} \underline{j} + \begin{vmatrix} 1 & 0 \\ 1 & 0 \end{vmatrix} \underline{k} = -2\underline{j}$.

33. The volume of the parallelepiped spanned by $a_1 \underline{i} + a_2 \underline{j} + a_3 \underline{k}$, $b_1 \underline{i} + b_2 \underline{j} + b_3 \underline{k}$, and $c_1 \underline{i} + c_2 \underline{j} + c_3 \underline{k}$, is the absolute value of $\begin{vmatrix} a_1 & a_2 & a_3 \\ b_1 & b_2 & b_3 \\ c_1 & c_2 & c_3 \end{vmatrix}$. The vectors from $(1,1,2)$ to $(2,0,2)$, $(3,1,3)$, and $(2,2,-3)$ are $\underline{i} - \underline{j}$, $2\underline{i} + \underline{k}$, and $\underline{i} + \underline{j} - 5\underline{k}$, respectively. Thus, $\begin{vmatrix} 1 & -1 & 0 \\ 2 & 0 & 1 \\ 1 & 1 & -5 \end{vmatrix} = 1 \begin{vmatrix} 0 & 1 \\ 1 & -5 \end{vmatrix} + 1 \begin{vmatrix} 2 & 1 \\ 1 & -5 \end{vmatrix} + 0 \begin{vmatrix} 2 & 0 \\ 1 & 1 \end{vmatrix} = -1 - 11 + 0 = -12$, and so the volume is 12 .

37. Consider $\begin{vmatrix} a & b & c \\ d & e & f \\ g & h & i \end{vmatrix}$. It is $a\begin{vmatrix} e & f \\ h & i \end{vmatrix} - b\begin{vmatrix} d & f \\ g & i \end{vmatrix} + c\begin{vmatrix} d & e \\ g & h \end{vmatrix} = a(ei - fh) -$

b(di - fg) + c(dh - eg) . Now, interchange the first two columns and

the determinant becomes $b\begin{vmatrix} d & f \\ g & i \end{vmatrix} - a\begin{vmatrix} e & f \\ h & i \end{vmatrix} + c\begin{vmatrix} e & d \\ h & g \end{vmatrix} = b(di - fg) -$

a(ei - fh) + c(eg - dh) , as required.

41.

Pick an arbitrary
right-handed set of
vectors and apply the
right-hand rule.

45. If $x = \begin{vmatrix} d_1 & b_1 & c_1 \\ d_2 & b_2 & c_2 \\ d_3 & b_3 & c_3 \end{vmatrix}/D$, $y = \begin{vmatrix} a_1 & d_1 & c_1 \\ a_2 & d_2 & c_2 \\ a_3 & d_3 & c_3 \end{vmatrix}/D$, and $z = \begin{vmatrix} a_1 & b_1 & d_1 \\ a_2 & b_2 & d_2 \\ a_3 & b_3 & d_3 \end{vmatrix}/D$,

then $a_1 x + b_1 y + c_1 z = (1/D)[a_1(b_2 c_3 d_1 + b_1 c_2 d_3 + b_3 c_1 d_2 - b_2 c_1 d_3 -$

$b_3 c_2 d_1 - b_1 c_3 d_2) + b_1(a_1 c_3 d_2 + a_3 c_2 d_1 + a_2 c_1 d_3 - a_3 c_1 d_2 - a_1 c_2 d_3 -$

$a_2 c_3 d_1) + c_1(a_1 b_2 d_3 + a_3 b_1 d_2 + a_2 b_3 d_1 - a_3 b_2 d_1 - a_1 b_3 d_2 - a_2 b_1 d_3)] =$

$(d_1/D)(a_1 b_2 c_3 - a_1 b_3 c_2 + a_3 b_1 c_2 - a_2 b_1 c_3 + a_2 b_3 c_1 - a_3 b_2 c_1) = d_1$.

Similarly, we can show that $a_2 x + b_2 y + c_2 z = d_2$ and $a_3 x + b_3 y +$

$c_3 z = d_3$.

49. Expanding along the first column, we get $\begin{vmatrix} a & d & g \\ b & e & h \\ c & g & i \end{vmatrix} = a\begin{vmatrix} e & h \\ f & i \end{vmatrix} - b\begin{vmatrix} d & g \\ f & i \end{vmatrix} +$

$c\begin{vmatrix} d & g \\ e & h \end{vmatrix}$. Expanding along the first row, we get $\begin{vmatrix} a & b & c \\ d & e & f \\ g & h & i \end{vmatrix} = a\begin{vmatrix} e & f \\ h & i \end{vmatrix} -$

$b\begin{vmatrix} d & f \\ g & i \end{vmatrix} + c\begin{vmatrix} d & e \\ g & h \end{vmatrix}$, which is $a\begin{vmatrix} e & h \\ f & i \end{vmatrix} - b\begin{vmatrix} d & g \\ f & i \end{vmatrix} + c\begin{vmatrix} d & g \\ e & h \end{vmatrix}$, by the result of

Example 2(b). Thus, the determinant of the transpose of a 3 × 3 matrix

is equal to the determinant of the original matrix.

53. Since all of the components of the third row of Exercise 18 are all zero

except for one, we expand along that row. This yields $\begin{vmatrix} 1 & 1 & 2 \\ 2 & -1 & 1 \\ 1 & 0 & 0 \end{vmatrix}$ =

$1 \begin{vmatrix} 1 & 2 \\ -1 & 1 \end{vmatrix} + 0 + 0 = 1 + 2 = 3$.

For Exercise 19, we apply the result of Exercise 50. The objective
is to obtain a determinant with one row or column whose entries are all
zero except for one. Add the third column to the second column to get

$\begin{vmatrix} 1 & 2 & 3 \\ -1 & -1 & 2 \\ 0 & 1 & -1 \end{vmatrix} = \begin{vmatrix} 1 & 5 & 3 \\ -1 & 1 & 2 \\ 0 & 0 & -1 \end{vmatrix} = 0 + 0 - 1 \begin{vmatrix} 1 & 5 \\ -1 & 1 \end{vmatrix} = -1(1 + 5) = -6$.

SECTION QUIZ

1. (a) Compute $\begin{vmatrix} 2 & 1 & 0 \\ 0 & 1 & 1 \\ 1 & 3 & 2 \end{vmatrix}$. What is the geometric interpretation of

this determinant? Is this a right-handed system?

(b) Compute $\begin{vmatrix} -1 & 3 & 2 \\ 2 & -6 & -4 \\ 3 & 0 & 1 \end{vmatrix}$. What is the geometric interpretation of

this determinant?

2. $\begin{vmatrix} 5 & -3 & 1 \\ 2 & 0 & 2 \\ 8 & -1 & 3 \end{vmatrix} = 5 \begin{vmatrix} 0 & 2 \\ -1 & 3 \end{vmatrix} - 3 \begin{vmatrix} 2 & 2 \\ 8 & 3 \end{vmatrix} + 1 \begin{vmatrix} 2 & 0 \\ 8 & -1 \end{vmatrix} = 5(-2) - 3(-10) + 1(-2) = 18$.

Two errors were made in the calculation. What are they?

3. Is the determinant in Question 2 equal to $-2 \begin{vmatrix} -3 & 1 \\ -1 & 3 \end{vmatrix} - 2 \begin{vmatrix} 5 & -3 \\ 8 & -1 \end{vmatrix}$? Explain

why or why not.

4.   Compute the following cross products:

(a) $(3\underline{i} + 3\underline{j}) \times (\underline{i} + 2\underline{j})$

(b) $(\underline{i} + \underline{j}) \times (-\underline{k})$

(c) $(2\underline{i} + 4\underline{j} + 6\underline{k}) \times (\underline{i} + 2\underline{j} + 3\underline{k})$

5.   Having found a key to the executive washroom, you decide to find out why the executive washroom is so special. Inside, you find silkworms busily manufacturing silk towellettes to fill a crystal parallelepiped. Vertices of the parallelepiped located adjacent to (0,1,1) are (1,3,2) , (2,5,4) , and (-1,-3,0) . Use determinants to compute what volume of silk is necessary to fill the parallelepiped crystal.

ANSWERS TO PREREQUISITE QUIZ

1. $\underline{k}$

2. $3\underline{i} - 2\underline{j} + 2\underline{k}$

ANSWERS TO SECTION QUIZ

1.   (a) -1 ; the parallelepiped spanned by the vectors (2,1,0) ,
           (0,1,1) , and (1,3,2) has volume 1 ; it is a left-handed
           system.

      (b) 0 ; all three vectors lie in a single plane.

2. $-3\begin{vmatrix} 2 & 2 \\ 8 & 3 \end{vmatrix}$ should be $+3\begin{vmatrix} 2 & 2 \\ 8 & 3 \end{vmatrix}$; $\begin{vmatrix} 0 & 2 \\ -1 & 3 \end{vmatrix} = +2$, not -2 .

3.   Yes; this is the expansion across the second row.

4.   (a) $3\underline{k}$

      (b) $-\underline{i} + \underline{j}$

      (c) $\underline{0}$

5.   2

13.R Review Exercises for Chapter 13

SOLUTIONS TO EVERY OTHER ODD EXERCISE

1. By the addition of ordered pairs, $(3,2) + (-1,6) = (3 - 1, 2 + 6) = (2,8)$.

5. In terms of ordered triples, $3\underline{i} + 2\underline{j}$ corresponds to $(3,2,0)$ and
    $8\underline{i} - \underline{j} - \underline{k}$ corresponds to $(8,-1,-1)$. Thus, $(3,2,0) + (8,-1,-1) =$
    $(11,1,-1)$ corresponds to $11\underline{i} + \underline{j} - \underline{k}$.

9. By the definition of dot products, $(a_1\underline{i} + b_1\underline{j} + c_1\underline{k}) \cdot (a_2\underline{i} + b_2\underline{j} + c_2\underline{k}) =$
    $a_1 a_2 + b_1 b_2 + c_1 c_2$. Therefore, $(8\underline{i} + 3\underline{j} - \underline{k}) \cdot (\underline{i} - \underline{j} - \underline{k}) = 8(1) +$
    $3(-1) + (-1)(-1) = 6$.

13. $\underline{u} \times \underline{v} = (2\underline{i} + \underline{j}) \times \underline{k} = 2\underline{i} \times \underline{k} + \underline{j} \times \underline{k}$. Using the fact that $\underline{i} \times \underline{k} = -\underline{j}$
    and $\underline{j} \times \underline{k} = \underline{i}$, the cross product becomes $\underline{i} - 2\underline{j}$.

17. Since $\underline{v}_1 \times \underline{v}_2$ is orthogonal to both $\underline{v}_1$ and $\underline{v}_2$, a vector orthogonal
    to $3\underline{i} + 2\underline{k}$ and $\underline{j} - \underline{k}$ is $(3\underline{i} + 2\underline{k}) \times (\underline{j} - \underline{k}) = 3\underline{i} \times \underline{j} + 2\underline{k} \times \underline{j} -$
    $3\underline{i} \times \underline{k} - 2\underline{k} \times \underline{k} = 3\underline{k} - 2\underline{i} + 3\underline{j} - \underline{0}$. Its length is $\sqrt{9 + 4 + 9} = \sqrt{22}$.
    Thus, $\underline{u} = (-2/\sqrt{22})\underline{i} + (3/\sqrt{22})\underline{j} + (3/\sqrt{22})\underline{k}$.

21. (a)

The vector joining $(-2,0)$ to $(4,6)$ has
components $(4 + 2, 6 - 0) = (6,6)$.

(b) The vector joining $(-2,0)$ to $(1,1)$ is $3\underline{i} + \underline{j}$. Adding $\underline{v}$
yields $9\underline{i} + 7\underline{j}$.

25.

Geometrically, the bird's velocity vector is
40 at $45°$ is $(40/\sqrt{2})\underline{i} + (40/\sqrt{2})\underline{j}$. The
wind's velocity vector is $-15\underline{j}$. The speed
of the bird relative to the earth's surface
is determined by the sum of the two vectors: $(40/\sqrt{2})\underline{i} + (40/\sqrt{2} - 15)\underline{j}$.
The speed is the length: $(800 + 800 - 1200/\sqrt{2} + 225)^{1/2} = (1825 - 600\sqrt{2})^{1/2}$
≈ 31.25 km/hr.

29. The direction of the line is $\underline{i} + \underline{j} + \underline{k}$, so it is $x = 1 + t$, $y = 1 + t$, $z = 2 + t$.

33. The vectors from $(1,1,2)$ to $(2,2,3)$ and $(0,0,0)$ are $\underline{i} + \underline{j} + \underline{k}$ and $-\underline{i} - \underline{j} - 2\underline{k}$, respectively. The normal to the plane is given by

the cross product: $\begin{vmatrix} \underline{i} & \underline{j} & \underline{k} \\ 1 & 1 & 1 \\ -1 & -1 & -2 \end{vmatrix} = -\underline{i} + \underline{j}$. Thus, the plane is $-x +$ $y + D = 0$. Substituting $(0,0,0)$ yields $D = 0$, so the plane is $-x + y = 0$.

37. The orthogonal line has direction $-\underline{i} + \underline{j}$, obtained from the coefficients of x , y , and z . Thus, the line is $x = -t$, $y = t$, $z = 3$.

41. From the coefficients, the orthogonal vector has direction $\underline{i} - 6\underline{j} + \underline{k}$. Its length is $\sqrt{1 + 36 + 1} = \sqrt{38}$. Thus, the desired unit vector is $(1/\sqrt{38})\underline{i} - (6/\sqrt{38})\underline{j} + (1/\sqrt{38})\underline{k}$.

45. Let the desired vector be $\underline{u} = a\underline{i} + b\underline{j} + c\underline{k}$. Then, $\underline{u} \cdot \underline{i} = \cos 30° = \sqrt{3}/2 = a$. Also, $\underline{u} \cdot \underline{j} = b = \underline{u} \cdot \underline{k} = c$. Since $\|\underline{u}\| = 1$, $a^2 + b^2 + c^2 = 1$. Substitute for a and b to get $3/4 + 2c^2 = 1$, so $c = 1/2\sqrt{2}$. Therefore, $\underline{u} = (\sqrt{3}/2)\underline{i} + (1/2\sqrt{2})\underline{j} + (1/2\sqrt{2})\underline{k}$.

49. 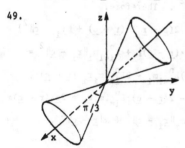 This is a (double) cone with its vertex at the origin, its axis along the x-axis, and an apical angle of $60°$.

53.

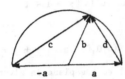

From the diagram, we see that $-\underline{a} + \underline{c} = \underline{b}$
and $\underline{a} + \underline{d} = \underline{b}$. Thus, $(-\underline{a} + \underline{c})\cdot(\underline{a} + \underline{d}) =$
$\underline{b}\cdot\underline{b} = \|\underline{b}\|^2 = -\underline{a}\cdot\underline{a} + \underline{c}\cdot\underline{a} - \underline{a}\cdot\underline{d} + \underline{c}\cdot\underline{d} =$
$-\|\underline{a}\|^2 + \underline{a}\cdot(\underline{c} - \underline{d}) + \underline{c}\cdot\underline{d}$. Since $\|\underline{b}\| =$
$\|\underline{a}\|$, we have $2\|\underline{a}\|^2 = \underline{a}\cdot(\underline{c} - \underline{d}) + \underline{c}\cdot\underline{d}$. Now, $\underline{c} - \underline{d} = 2\underline{a}$, so
$2\|\underline{a}\|^2 = \underline{a}\cdot(2\underline{a}) + \underline{c}\cdot\underline{d} = 2\|\underline{a}\|^2 + \underline{c}\cdot\underline{d}$, i.e., $\underline{c}\cdot\underline{d} = 0$. Therefore $\underline{c}$
is perpendicular to $\underline{d}$.

57. The determinant $\begin{vmatrix} a & b \\ c & d \end{vmatrix}$ is $ad - bc$, so $\begin{vmatrix} -1 & -1 \\ 2 & 1 \end{vmatrix} = (-1)(1) -$
$(-1)(2) = -1 + 2 = 1$.

61. Expanding across the first row yields $\begin{vmatrix} 1 & 1 & 1 \\ 2 & 2 & 2 \\ 3 & 3 & 3 \end{vmatrix} = 1\begin{vmatrix} 2 & 2 \\ 3 & 3 \end{vmatrix} - 1\begin{vmatrix} 2 & 2 \\ 3 & 3 \end{vmatrix} +$
$1\begin{vmatrix} 2 & 2 \\ 3 & 3 \end{vmatrix} = 0$.

65. The volume is the absolute value of $\begin{vmatrix} 1 & -1 & -1 \\ 2 & 1 & -5 \\ 8/3 & -1 & 1/2 \end{vmatrix} = 1\begin{vmatrix} 1 & -5 \\ -1 & 1/2 \end{vmatrix} +$
$1\begin{vmatrix} 2 & -5 \\ 8/3 & 1/2 \end{vmatrix} - 1\begin{vmatrix} 2 & 1 \\ 8/3 & -1 \end{vmatrix} = -9/2 + 40/3 + 14/3 = 29/2$.

69. $\|\underline{r} - \underline{r}_i\|^2 = (\underline{r} - \underline{r}_i)\cdot(\underline{r} - \underline{r}_i) = ((\underline{r} - \underline{c}) + (\underline{c} - \underline{r}_i))\cdot((\underline{r} - \underline{c}) +$
$(\underline{c} - \underline{r}_i)) = (\underline{r} - \underline{c})\cdot(\underline{r} - \underline{c}) + 2(\underline{c} - \underline{r}_i)\cdot(\underline{r} - \underline{c}) + (\underline{c} - \underline{r}_i)\cdot(\underline{c} - \underline{r}_i) =$
$\|\underline{r} - \underline{c}\|^2 + 2(\underline{c} - \underline{r}_i)\cdot(\underline{r} - \underline{c}) + \|\underline{c} - \underline{r}_i\|^2$. Therefore,
$S = \Sigma_{i=1}^n m_i \|\underline{r} - \underline{r}_i\|^2 = \Sigma_{i=1}^n m_i \{\|\underline{r} - \underline{c}\|^2 + 2(\underline{c} - \underline{r}_i)\cdot(\underline{r} - \underline{c}) + \|\underline{r}_i - \underline{c}\|^2\} =$
$\|\underline{r} - \underline{c}\|^2 \Sigma_{i=1}^n m_i + 2\Sigma_{i=1}^n m_i [\underline{c}\cdot(\underline{r} - \underline{c}) - \underline{r}_i\cdot(\underline{r} - \underline{c})] + \Sigma_{i=1}^n m_i \|\underline{r}_i - \underline{c}\|^2 =$
$\Sigma_{i=1}^n m_i \|\underline{r}_i - \underline{c}\|^2 + m\|\underline{r} - \underline{c}\|^2 + 2\underline{c}\cdot(\underline{r} - \underline{c})\Sigma_{i=1}^n m_i - 2\Sigma_{i=1}^n m_i \underline{r}_i\cdot(\underline{r} - \underline{c})$.
The last two terms become $2\underline{c}\cdot(\underline{r} - \underline{c})m - 2(\underline{r} - \underline{c})\cdot\Sigma_{i=1}^n m_i \underline{r}_i = 2\underline{c}\cdot(\underline{r} - \underline{c})m -$
$2(\underline{r} - \underline{c})\cdot m\underline{c} = 0$. Therefore, $S = \Sigma_{i=1}^n m_i \|\underline{r}_i - \underline{c}\|^2 + m\|\underline{r} - \underline{c}\|^2$.

73. The objective is to obtain a row or column whose entries are all zero except for one. This is done by adding a multiple of one row or column to another. In this case, we add row 2 to row 1 and then add $-1/2$ of column 1 to column 2. This yields $\begin{vmatrix} 6 & 2 & -3 \\ 2 & 2 & 3 \\ 4 & 8 & -1 \end{vmatrix} = \begin{vmatrix} 8 & 4 & 0 \\ 2 & 2 & 3 \\ 4 & 8 & -1 \end{vmatrix} = \begin{vmatrix} 8 & 0 & 0 \\ 2 & 1 & 3 \\ 4 & 6 & -1 \end{vmatrix} = 8 \begin{vmatrix} 1 & 3 \\ 6 & -1 \end{vmatrix} = 8(-1 - 18) = -162$. We expanded along the first row.

77. By Example 8, Section 13.6, $(\underline{v} \times \underline{j}) \cdot \underline{k} = (\underline{j} \times \underline{k}) \cdot \underline{v} = \underline{i} \cdot \underline{v} = 0$. Thus, $\underline{v}$ is orthogonal to $\underline{i}$.

81. Let $\underline{u} = a_1 \underline{i} + b_1 \underline{j} + c_1 \underline{k}$, $\underline{v} = a_2 \underline{i} + b_2 \underline{j} + c_2 \underline{k}$, and $\underline{w} = a_3 \underline{i} + b_3 \underline{j} + c_3 \underline{k} \neq \underline{0}$.

   (a) $\underline{v} \cdot \underline{w} = \|\underline{v}\| \|\underline{w}\| \cos \theta$ by definition. Since $\|\underline{w}\| \neq 0$ and there exist values of θ such that $\cos \theta \neq 0$, then $\underline{v} \cdot \underline{w} = 0$ for all $\underline{w}$. This implies that $\|\underline{v}\| = 0 = \sqrt{a_2^2 + b_2^2 + c_2^2}$. Therefore, $a_2 = b_2 = c_2 = 0$, and $\underline{v} = \underline{0}$.

   (b) $\underline{u} \cdot \underline{w} = \underline{v} \cdot \underline{w}$ implies $\underline{u} \cdot \underline{w} - \underline{v} \cdot \underline{w} = 0$. Therefore, $(\underline{u} - \underline{v}) \cdot \underline{w} = 0$. From part (a), we conclude that $\underline{u} - \underline{v} = \underline{0}$, so $\underline{u} = \underline{v}$.

   (c) We have $\underline{v} \cdot \underline{i} = a_2 = \underline{v} \cdot \underline{j} = b_2 = \underline{v} \cdot \underline{k} = c_2 = 0$, so $\underline{v} = \underline{0}$.

   (d) We have $\underline{u} \cdot \underline{i} = a_1 = \underline{v} \cdot \underline{i} = a_2$; $\underline{u} \cdot \underline{j} = b_1 = \underline{v} \cdot \underline{j} = b_2$; and $\underline{u} \cdot \underline{k} = c_1 = \underline{v} \cdot \underline{k} = c_2$. Since their components are equal, $\underline{u} = \underline{v}$.

85. In each case, $rP + sQ = r(1,2) + s(\pi,2\pi) = (r + \pi s, 2r + 2\pi s)$.
Therefore, all such points must lie on the line $y = 2x$.

(a)

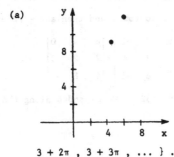

Here, the x-coordinate is $x = r + \pi s$, where r and s are any combination of positive integers. Thus, x is in the set $\{1 + \pi ,$ $1 + 2\pi , 1 + 3\pi , \ldots , 2 + \pi ,$ $2 + 2\pi , 2 + 3\pi , \ldots , 3 + \pi ,$ $3 + 2\pi , 3 + 3\pi , \ldots \}$.

(b)

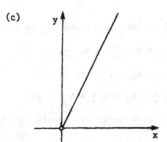

Here, the x-coordinate is $x = r + \pi s$, where r and s are any combination of integers. Thus, x is in the set $\{0 , \pm\pi , \pm2\pi , \pm3\pi ,$ $\ldots , \pm1 , \pm1 \pm \pi, \pm1 \pm 2\pi , \pm1 \pm$ $3\pi , \ldots , \pm2 , \pm2 \pm \pi , \pm2 \pm 2\pi ,$ $\pm2 \pm 3\pi , \ldots , \pm3 , \pm3 \pm \pi , \pm3 \pm 2\pi , \pm3 \pm 3\pi , \ldots \}$.

(c)

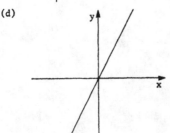

Here, the x-coordinate is $x = r + \pi s$, where r and s are any combination of positive real numbers. Thus, x is any positive real number.

(d)

Here, the x-coordinate is $x = r + \pi s$, where r and s are any combination of real numbers. Thus, x is any real number.

TEST FOR CHAPTER 13

1. True or false.

   (a) Any non-zero vector $\underline{v}$, dotted with the zero vector, is the
       zero vector.

   (b) Two vectors originating from different points in space are equal
       if both have the same magnitude and the same direction.

   (c) Lines in space which do not lie in parallel planes must intersect.

   (d) There is exactly one vector in space which can not be normalized
       to a unit vector.

   (e) Both the cross product and the dot product may be used to calculate
       the angle between two vectors.

2. A plane P contains the origin and has a normal vector $2\underline{i} + \underline{j} - \underline{k}$.
   Find a plane perpendicular to P which contains the line $x = 5 + 2t$,
   $y = 3 - t$, $z = -10 + 5t$.

3. Perform the following calculations:

   (a) $(\underline{i} + \underline{j} - \underline{k}) \cdot (2\underline{i} - 3\underline{j})$

   (b) $3(2,1) + (2,0)$

   (c) $(2\underline{i} - \underline{k}) \times (\underline{i} + \underline{j})$

   (d) $\begin{vmatrix} 2 & 0 & 1 \\ 1 & 0 & 5 \\ 0 & -1 & 3 \end{vmatrix}$

4. (a) State how the cross product $\underline{u} \times \underline{v}$ is related to the angle θ
       between $\underline{u}$ and $\underline{v}$.

   (b) State how the dot product $\underline{u} \cdot \underline{v}$ is related to the angle θ
       between $\underline{u}$ and $\underline{v}$.

   (c) Find a unit vector in the plane which is orthogonal to $3\underline{i} - 2\underline{j}$.

5. Let $(1,1,1)$, $(2,1,0)$, and $(1,-1,3)$ be the vertices of a triangle.

(a) What is the area of the triangle?

(b) What is the equation of the plane passing through the three points?

6. Find a set of parametric equations which describe the line passing through the point $(2,6,7)$ and perpendicular to the plane

$4x + 3y - 4z + 2 = 0$.

7. Let L be the line $(-2,0,1) + t(3,2,2)$ and let M be the line

$(3,5,4) + t(1,-1,1)$.

(a) Where do the two lines intersect?

(b) Find the equation of a plane which is parallel to the two lines and exactly one unit distance away from the lines.

8. Let P , Q , and R be the points $(1,0,0)$, $(0,2,3)$, and $(-1,0,-1)$, respectively.

(a) Does $\overrightarrow{PQ} - \overrightarrow{PR}$ equal $\overrightarrow{QR}$ or $\overrightarrow{RQ}$?

(b) What is the orthogonal projection of $\overrightarrow{PQ}$ upon $\overrightarrow{PR}$?

(c) What is the distance from R to the line passing through P and Q ?

9.

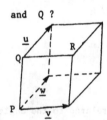

In the parallelepiped shown at the left, $\underline{u} = -2\underline{i} + \underline{j} + 2\underline{k}$, $\underline{v} = 3\underline{j} - \underline{k}$, $P = (0,0,0)$, and $Q = (1,1,1)$. The parallelepiped is not drawn to scale.

(a) What are the coordinates of R ?

(b) What is the distance from R to the plane spanned by $\underline{v}$ and $\underline{w}$?

(c) Compute the volume of the parallelepiped.

10. On a brisk first day of Spring, a young man has just finished waxing his car for his date later that evening. Unfortunately, a bird dropping comes out of the sky from $(-2,0,10)$ and falls with velocity $-2\underline{k}$. A strong gust of wind suddenly begins blowing with velocity $\underline{i}$. If the corners of the car are located at $(2.5 \pm 0.3, 0 \pm 0.1, 0.05 \pm 0.05)$ and the young man is standing at $(2.9,0,0)$, will the bird dropping ruin the wax job? Explain your answer.

ANSWERS TO CHAPTER TEST

1. (a) False; dot products are scalars.

   (b) True

   (c) False; consider the x-axis in the xy-plane and any line in the plane $x = 1$ which does not intersect the x-axis.

   (d) True; the zero vector.

   (e) True

2. $-x + 3y + z + 6 = 0$

3. (a) -1

   (b) $(8,3)$

   (c) $\underline{i} - \underline{j} + 2\underline{k}$

   (d) 9

4. (a) $\|\underline{u} \times \underline{v}\| = \|\underline{u}\|\|\underline{v}\| \sin \theta$

   (b) $\underline{u} \cdot \underline{v} = \|\underline{u}\|\|\underline{v}\| \cos \theta$

   (c) $(2\underline{i} + 3\underline{j})\sqrt{13}$ or $(-2\underline{i} - 3\underline{j})/\sqrt{13}$

5. (a) $\sqrt{3}$

   (b) $-x - y - z + 3 = 0$

6. $x = 2 + 4t$, $y = 6 + 3t$, $z = 7 - 4t$

7.   (a) $(4,4,5)$

   (b) $4x - y - 5z = -13 + \sqrt{42}$ or $4x - y - 5z = -13 - \sqrt{42}$

8.   (a) $\overrightarrow{RQ}$

   (b) $(2\underline{i} - \underline{k})/5$

   (c) $\sqrt{69/14}$

9.   (a) $(1,4,0)$

   (b) $15/\sqrt{89}$

   (c) 15

10. No; it travels along the line $(-2,0,10) + t(1,0,-2)$, so it will not
     hit the car. (The young man will probably be hit in the forehead.)

CHAPTER 14

CURVES AND SURFACES

14.1 The Conic Sections

PREREQUISITES

1. Recall the concept of an asymptote (Section 3.4).
2. Recall how to sketch a parabola and recognize its equation
   (Section R.5).

PREREQUISITE QUIZ

1. Find the horizontal and vertical asymptotes of $f(x) = (x + 2)/(x - 3)$.
2. Which are the following equations represent a parabola?

   (a) $y^2 + x^2 = 4y$

   (b) $x = y^2 - 3$

   (c) $-x^2 - 2 = y$

3. Sketch the graph of any parabolas whose equation is given in Question 2.

GOALS

1. Be able to write equations of ellipses, parabolas, and hyperbolas in
   standard form and sketch them.

STUDY HINTS

1. <u>Ellipse</u>. The standard form is $x^2/a^2 + y^2/b^2 = 1$. In an ellipse, the
    sum of the distances from two foci to a point on the ellipse is constant.
    You often don't need to be able to find the foci, but you should be able
    to convert to the standard form and use the equation to sketch an ellipse.
    The x-intercepts are $\pm a$ and the y-intercepts are $\pm b$, found by sub-
    stituting $y = 0$ and $x = 0$, respectively.

2. <u>Hyperbola</u>. The standard form is either $x^2/a^2 - y^2/b^2 = 1$ or $y^2/a^2 -$
    $x^2/b^2 = 1$. Here, the difference of the distances from two foci to a
    point on the hyperbola is constant. Again, you probably don't always
    need to find the foci, but you should be able to convert to the stan-
    dard form and sketch a hyperbola. By drawing a rectangle with sides at
    $x = \pm a$ and $y = \pm b$, the diagonals of the rectangle represent the
    asymptotes $y = \pm(b/a)x$. The intercepts are either $x = \pm a$ or $y =$
    $\pm b$. Substitution determines which are the intercepts.

3. <u>Circles</u>. A circle is a special ellipse in which $a = b$.

4. <u>Parabolas</u>. Your main concern is to be able to sketch a parabola. It
    is less important to be able to determine the directrix and the focus.

5. <u>Circles and parabolas</u>. Look back over Section R.5 and make sure you
    understand that material in the present context.

SOLUTIONS TO EVERY OTHER ODD EXERCISE

1.   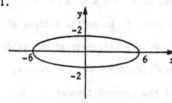   Dividing through by 36, the equation
becomes $x^2/(6)^2 + y^2/(2)^2 = 1$. Since
$2 < 6$, the foci are $(\pm c,0)$, where
$c = \sqrt{6^2 - 2^2}$, i.e., the foci are
$(\pm\sqrt{32},0)$. The intercepts are $(\pm 6,0)$
and $(0,\pm 2)$.

5.   Dividing through by 4 , $y^2 - x^2 = 2$ is
equivalent to $y^2/(\sqrt{2})^2 - x^2/(\sqrt{2})^2 = 1$.
Thus, the foci are $(0,\pm c)$, where $c = \sqrt{(\sqrt{2})^2 + (\sqrt{2})^2}$, i.e., the foci are
$(0,\pm 2)$. The asymptotes are $y = \pm(\sqrt{2}/\sqrt{2})x = \pm x$, and the intercepts are
at $(0,\pm\sqrt{2})$.

9.   From the given information, we have $c = 4$ and $a = 1/4c = 1/16$.
Therefore, the equation of the parabola is $y = x^2/16$.

13. The equation has the form $x = by^2$, so the focus is at $(c,0)$, where
$c = 1/4b = 1/4$. Thus, the focus is $(1/4,0)$ and the directrix is
$x = -1/4$.

17. Since the vertex is at $(0,0)$, try the general form $y = ax^2$. Since
the parabola passes through $(2,1)$, $y = ax^2$ becomes $1 = 4a$. Thus,
$a = 1/4$ and the equation is $y = x^2/4$.

21. Use the method of Example 6. The equation of the parabola is $y = ax^2$.
Since $y = 0.3$ when $x = 0.4$, we get $a = 0.3/0.16 = 15/8$. The
focus is at $(0,c)$, where $a = 1/4c$, so $c = 1/4a = 1/(15/2) = 2/15$.
Thus, the light source should be placed on the axis, $2/15$ meters from
the mirror.

25.

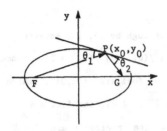

Consider the equation of the ellipse in standard form with $b < a$: $x^2/a^2 + y^2/b^2 = 1$. Let P be on the ellipse at (x_0, y_0) . Implicit differentiation yields $2x/a^2 + (2y/b^2)(dy/dx) = 0$. Therefore, the slope of the tangent line at P is

$dy/dx = -b^2 x_0/a^2 y_0$. Since the foci are located at $(\pm c, 0)$, we have $\vec{FP} = (x_0 + c)\underline{i} + y_0 \underline{j}$ and $\vec{PG} = (c - x_0)\underline{i} - y_0 \underline{j}$. Also, the direction of the tangent line is $\vec{d} = -a^2 y_0 \underline{i} + b^2 x_0 \underline{j}$. Now, by the definition of the dot product, $\vec{d} \cdot \vec{FP} = (\cos \theta_1)\|\vec{d}\|\|\vec{FP}\| = -a^2 y_0(x_0 + c) + b^2 x_0 y_0 = (b^2 - a^2)x_0 y_0 - a^2 y_0 c = -(c^2 x_0 y_0 + a^2 y_0 c) = -cy_0(cx_0 + a^2)$ and $\vec{d} \cdot \vec{PG} = (\cos \theta_2)\|\vec{d}\|\|\vec{PG}\| = a^2 y_0(c - x_0) + b^2 x_0 y_0 = (b^2 - a^2)x_0 y_0 + a^2 y_0 c = -(c^2 x_0 y_0 - a^2 y_0 c) = -cy_0(cx_0 - a^2)$. We used the fact that $c^2 = a^2 - b^2$. Rearrangement gives us $\cos \theta_1 = -cy_0(cx_0 + a^2)/\|\vec{d}\|\|\vec{FP}\|$ and $\cos \theta_2 = -cy_0(cx_0 - a^2)/\|\vec{d}\|\|\vec{PG}\|$.

From the figure, if the angle from $\vec{d}$ to $\vec{PG}$ is $-\phi$, then the angle from $\vec{d}$ to $\vec{FP}$ is $\pi - \phi$. Thus, we need to show that $\cos \theta_1/\cos \theta_2 = -1 = [-cy_0(cx_0 + a^2)/\|\vec{d}\|\|\vec{FP}\|]/[-cy_0(cx_0 - a^2)/ \|\vec{d}\|\|\vec{PG}\|]$. This simplifies to $(cx_0 + a^2)/\|\vec{FP}\| = -(cx_0 - a^2)/\|\vec{PG}\|$. Use the fact that $\|\vec{FP}\| = \sqrt{(x_0 + c)^2 + y_0^2}$ and $\|\vec{PG}\| = \sqrt{(c - x_0)^2 + y_0^2}$.

Rearrange the equation and square to get $\|\vec{PG}\|^2(cx_0 + a^2)^2 = \|\vec{FP}\|^2(cx_0 - a^2)^2$ or $((c - x_0)^2 + y_0^2)(cx_0 + a^2)^2 = ((x_0 + c)^2 + y_0^2) \times (cx_0 - a^2)^2$. Multiplying out, cancelling like terms, dividing by four, and rearrangement yields $-cx_0^3 a^2 - c^3 x_0 a^2 - cx_0 a^2 y_0^2 = c^3 x_0^3 + cx_0 a^4$, i.e., $-cx_0 a^2(x_0^2 + c^2 + y_0^2) = cx_0(c^2 x_0^2 + a^4)$, i.e., $-a^2(x_0^2 + c^2 + y_0^2) = c^2 x_0^2 + a^4$. Now, $x_0^2/a^2 + y_0^2/b^2 = 1$ rearranges to $y_0^2 = b^2 - x_0^2 b^2/a^2$ and using $c^2 = a^2 - b^2$ makes the left-hand side $-a^2(x_0^2 + c^2 + y_0^2) =$

25. (continued)

$$-a^2x_0^2 - a^2c^2 - a^2b^2 + x_0b^2 = x_0^2(b^2 - a^2) - a^2(c^2 + b^2) = -c^2x_0^2 - a^4 \, ,$$

which is the same as the right-hand side except for an irrelevant sign

change.

SECTION QUIZ

1. Classify the equations of the following conics:

    (a) $x^2 = 4(1 - y^2)$

    (b) $5y^2 - x = 0$

    (c) $4x^2 - y^2 = -4$

    (d) $4x^2 = 1 - 4y^2$

2. Which of the equations in Question 1 are functions of x ?

3. Sketch the conic sections described by the equations in Question 1.

4. What is the equation of the ellipse passing through the points (-5,0) ,

    (0,-7) , (5,0) , and (0,7) ?

5. Mother and father racoon were raiding separate campgrounds, looking for

    food to feed their family. Being expert food thieves, the two racoons

    had their methods synchronized. As they were breaking into separate

    backpacks, a hunter in the distance sighted a duck. A shot sent both

    racoons scurrying back into the wilderness without any food.

    (a) Suppose the campgrounds are separated by 900 m and the sound travels

        at 300 m/sec . If one racoon heard the shot 1 second before

        the other, describe all points where the hunter could possibly have

        been. [Hint: Let the campsites represent foci located at (±450,0) .]

    (b) Sketch the graph of the expression found in (a).

ANSWERS TO PREREQUISITE QUIZ

1. Horizontal asymptote: $y = 1$; vertical asymptote: $x = 3$

2. b and c

3. (b)

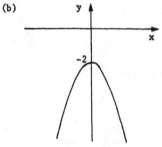

(c)

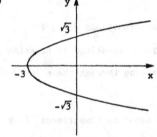

ANSWERS TO SECTION QUIZ

1. (a) Ellipse

    (b) Parabola

    (c) Hyperbola

    (d) Circle

2. None of them

3.   (a)

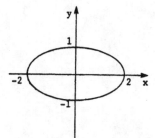

(b)

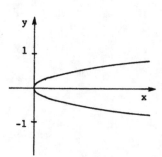

(c)

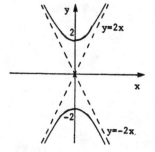

(d)

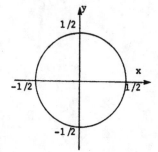

4. $x^2/25 + y^2/49 = 1$

5.   (a) $x^2/22500 - y^2/180000 = 1$

     (b)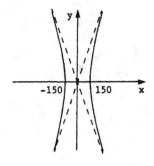

14.2 Translation and Rotation of Axes

PREREQUISITES

1. Recall how to find equations of shifted parabolas (Section R.5).

2. Recall the general equations for the conic sections (Sections R.5 and
   14.1).

PREREQUISITE QUIZ

1. What kind of conic section is represented by each of the following
   equations?

   (a) $x^2 + y^2 = 4$

   (b) $2x^2 + 3y^2 = 6$

   (c) $y^2 - x^2 = 2$

   (d) $y = x^2 + 2$

2. What is the vertex of the parabola with equation $y^2 + 2y + 1 = x - 1$?

GOALS

1. Be able to translate and rotate axes for the purpose of graphing conic
   sections.

STUDY HINTS

1. Translating axes. Translation of axes is called for if linear terms in
   x and y appear, but no cross-term xy appears. For purposes of
   sketching, let $X = x - p$ and $Y = y - q$. This allows you to write
   equations in standard form. Essentially, all that is being done is a
   shifting of the origin. It is a good idea to substitute a point into
   the original equation to determine if the origin was shifted in the
   correct direction.

2. <u>Rotating axes</u>. If the term xy is present, rotating axes can get rid
    of the term. The equations used to change (x,y) to (X,Y) are given
    in formula (5), p. 705. You should either remember these formulas or
    learn to derive them. To convert back to (x,y) from (X,Y) , use
    the angle $-\alpha$. Although these formulas should be known, it is a good
    idea to consult your instructor to see if you will be tested on them.

3. <u>General conics</u>. $Ax^2 + Bxy + Cy^2 + Dx + Ey + F = 0$ is the general equa-
    tion of a conic. By computing $B^2 - 4AC$, you should know that the
    graph (provided it exists) is an ellipse if this quantity is less than
    zero, is a parabola if $B^2 - 4AC = 0$, and is a hyperbola if $B^2 - 4AC >$
    0 . An important formula is $\tan 2\alpha = B/(A - C)$; this determines the
    angle of rotation, α . Finally, substituting the expressions in (5)
    helps you to get the standard form.

SOLUTIONS TO EVERY OTHER ODD EXERCISE

1.

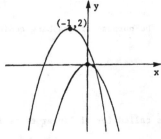

$y = -x^2$ is a parabola opening downward
and centered along the y-axis. Its vertex is (0,0) . $y - 2 = -(x + 1)^2$ is
the same parabola with the origin shifted
to (-1,2) .

5.

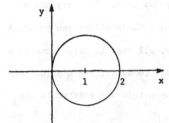

Complete the squares. $(x^2 - 2x + 1) + y^2 =$
$1 = (x - 1)^2 + y^2$ is a circle of radius 1
centered at (1,0) .

9.

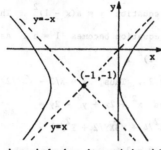

By completing the squares, the equation becomes $(x^2 + 2x + 1) - (y^2 + 2y + 1) = 1 + 1 - 1 = 1 = (x + 1)^2 - (y + 1)^2$. This has the form $X^2 - Y^2 = 1$, which is a hyperbola with asymptotes $Y = \pm X$ and intercepts $(\pm 1,0)$. The desired hyperbola has its origin shifted to $(-1,-1)$.

13. Use the method of Example 4. $\cos 15° \approx 0.97$ and $\sin 15° \approx 0.26$, so the transformation of coordinates is $x = 0.97X - 0.26Y$, $y = 0.26X + 0.97Y$, and $X = 0.97x + 0.26y$, $Y = -0.26x + 0.97y$.

17. We consider the equation $Ax^2 + Bxy + Cy^2 + Dx + Ey + F = 0$. It is an ellipse if $B^2 - 4AC < 0$, a hyperbola if $B^2 - 4AC > 0$, and a parabola if $B^2 - 4AC = 0$. In this case, $A = 19/4$, $B = 7\sqrt{3}/6$, and $C = 43/12$, so $B^2 - 4AC = 49/12 - 4(19/4)(43/12) = (49 - 19\cdot43)/12 < 0$. Thus, it is an ellipse.

21.

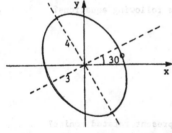

Follow the four-step procedure described in the box on p. 709. $\alpha = (1/2)\tan^{-1}(B/(A - C)) = (1/2)\tan^{-1}[(7\sqrt{3}/6)/(19/4 - 43/12)] = (1/2)\tan^{-1}(\sqrt{3}) = \pi/6$. Thus, $\cos(\pi/6) = \sqrt{3}/2$ and $\sin(\pi/6) = 1/2$, and $x = \sqrt{3}X/2 - Y/2$ and $y = X/2 + \sqrt{3}Y/2$; therefore, the given equation is $(19/4)(3X^2/4 - \sqrt{3}XY/2 + Y^2/4) + (43/12)(X^2/4 + \sqrt{3}XY/2 + 3Y^2/4) + (7\sqrt{3}/6)(\sqrt{3}X^2/4 + XY/2 - \sqrt{3}Y^2/4) - 48 = 16X^2/3 + 3Y^2 - 48 = 0$, i.e., $X^2/9 + Y^2/16 = 1$. This new equation is an ellipse with intercepts $(\pm 3,0)$ and $(0,\pm 4)$. Rotate this by $\pi/6$ to get the desired sketch.

25. Since the vertex is at $(1,0)$, try the equation $y = a(x - 1)^2$. The parabola passes through $(2,1)$, so the equation becomes $1 = a$. As a check, $(0,1)$ also satisfies $y = (x - 1)^2$.

29. Here, $\cos(\pi/3) = 1/2$ and $\sin(\pi/3) = \sqrt{3}/2$. Thus, $x = X/2 - \sqrt{3}Y/2$ and $y = \sqrt{3}X/2 + Y/2$. Substitution gives $x^2 + 2x + y^2 - 2y =$ $(X^2/4 - \sqrt{3}XY/2 + 3Y^2/4) + (X + \sqrt{3}Y) + (3X^2/4 + \sqrt{3}XY/2 + Y^2/4) -$ $(\sqrt{3}X + Y) = X^2 + Y^2 + (1 - \sqrt{3})X + (\sqrt{3} - 1)Y = 2$.

33. Since $B^2 - 4AC < 0$, the equation $Ax^2 + Bxy + Cy^2 = 1$ represents an ellipse. From formulas (9), this becomes $\bar{A}X^2 + \bar{C}Y^2 = 1$, where $A = \bar{A}\cos^2\alpha + \bar{C}\sin^2\alpha$, $C = \bar{A}\sin^2\alpha + \bar{C}\cos^2\alpha$, and $\tan 2\alpha = B/(A - C)$. $\bar{A}X^2 + \bar{C}Y^2 = 1$ represents an ellipse whose axes have length $1/\sqrt{\bar{A}}$ and $1/\sqrt{\bar{C}}$. Therefore, the area is $\pi/\sqrt{\bar{A}\bar{C}}$. But $AC - B^2/4 = \bar{A}\bar{C}$ from p. 707 and this gives the desired result.

SECTION QUIZ

1. What type of conics are described by the following equations?

    (a) $3u^2 - 5uv - 2v^2 + 8v - 3u = 0$

    (b) $y^2 + 4x^2 - 8x + 4y - 17 = 3$

    (c) $t^2 + 4xt + 4x^2 - 3t + 4x = 28$

    (d) $4x + xy - 2y^2 = 1$

2. Which of the equations in Question 1 represent rotated conics?

3. A circle of radius 3 is centered at $(-2,1)$ and it is rotated $30°$. What is the equation of the circle?

4. Sketch the graph of $(y^2 - x^2)/2 + \sqrt{3}xy = 0$.

5. Sketch the graph of the equation in Question 1(b).

6. Stranded in the middle of the Pacific Ocean on a rubber life raft, an
    unlucky gentleman has received the company of a shark. While worrying
    that the shark's sharp teeth will soon puncture the raft, our sea-faring
    friend notices that the shark is swimming in an elliptical path. He
    notes that if he is at the origin, the major axis is on the line
    $y = \sqrt{3}x$. The major axis has length 3 and the minor axis, whose length
    is 2 , intersects the major axis at $(1,\sqrt{3})$.

    (a) What is the equation of the ellipse?

    (b) Will the shark collide with the man? Explain.

ANSWERS TO PREREQUISITE QUIZ

1. (a) Circle

    (b) Ellipse

    (c) Hyperbola

    (d) Parabola

2. $(1,-1)$

ANSWERS TO SECTION QUIZ

1. (a) Hyperbola

    (b) Ellipse

    (c) Parabola

    (d) Hyperbola

2. a , c , and d

3. $x^2 + 4x + y^2 + 2y = 4$

4.

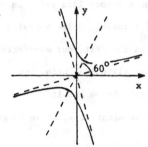

5.

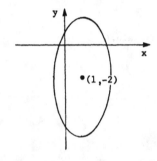

6.   (a) $85x^2 - 10\sqrt{3}xy + 21y^2 - 4x - 20\sqrt{3}y = 20$

   (b)   No; $(0,0)$ does not satisfy the equation in (a).

14.3 Functions, Graphs, and Level Surfaces

PREREQUISITES

1. Recall the concept of functions in one variable (Section R.6).

2. Recall how to graph functions of one variable (Section 3.4).

PREREQUISITE QUIZ

1. What is the domain of the following functions?

    (a) $f(t) = 2/(t - 3)$

    (b) $g(x) = \sqrt{x^2 - 3x + 2}$

2. Sketch the graphs of the functions in Question 1.

GOALS

1. Be able to define a function of several variables, a level curve, and
    a level surface.

2. Be able to recognize the equations of a plane or sphere and sketch the
    graphs.

3. Be able to apply the method of sections to sketch surfaces in space.

STUDY HINTS

1. Function defined. Recall that a function of one variable assigns a
    single, real value for each x in its domain. A function of two vari-
    ables assigns a single, real value for each (x,y) in its domain.
    Similarly, a function of three variables assigns a single, real value
    for each (x,y,z) in its domain.

2. Level curves and surfaces. A level curve is the intersection of $f(x,y)$
    and the plane z = c . You just sketch $f(x,y) = c$ in the xy-plane. A
    level surface is the surface which results if $f(x,y,z)$ is assigned a
    constant value. This concept is useful for sketching in three dimensions.

3. Sketching planes. Many of us are poor artists, and as a result, three-
    dimensional geometry may be frustrating due to this problem rather than
    a lack of mathematical understanding. Planes are most easily sketched
    by plotting three non-collinear points (usually on the coordinate
    axes) and then passing a plane through them.

    Another simple method is to set one of x , y , or z constant
    to obtain lines as level curves. Then use the method of sections
    described in Study Hint 6.

4. Spheres. Example 5(b) shows a surface of the form $x^2 + y^2 + z^2 = c^2$.
    This is a sphere of radius c centered at the origin. More generally,
    $(x - a)^2 + (y - b)^2 + (z - c)^2 = r^2$ is a sphere of radius r centered
    at (a,b,c) .

5. Cylinders. A surface is called a cylinder if x , y , or z is missing
    from its equation. A cylinder can be sketched by drawing the level curve
    in the plane where the missing variable is zero. Then move the curve
    along the axis of the missing variable, since no restrictions are placed
    upon the missing variable. Note that, in Example 6(b), the level curve
    was drawn in the plane z = 0 and then moved along the z-axis since z
    is missing in the equation. Cylinders may be classified according to
    their cross-sections as being elliptic, parabolic, circular, or
    hyperbolic.

6. Method of sections. The idea is to sketch level curves in planes
    parallel to the coordinate axes. Then, we put them together to obtain
    the three dimensional figure. If possible, try to look for a general
    pattern. For instance, in Example 7, we note that in the plane z = c ,
    the level curve is a circle of radius c is $z \geq 0$. Sometimes, these
    level curves are called traces.

SOLUTIONS TO EVERY OTHER ODD EXERCISE

1. The domain is (x,y) such that the denominator does not vanish. $x = 0$ is the y-axis. The domain is all points not on the y-axis; $f(1,0) = 0$; $f(1,1) = 1$.

5. The domain is (x,y) such that the denominator does not vanish. $x^2 + y^2 + z^2 = 1$ is the unit sphere. The domain is all points not on the unit sphere; $f(1,1,1) = 1$; $f(0,0,2) = -2/3$.

9.

    $f(x,y) = 1 - x - y$ represents a plane. The plane contains $(0,0,1)$, $(0,1,0)$, and $(1,0,0)$. Plot the three noncollinear points and sketch the plane which passes through all three.

13.

    $1 - x - y = 1$ implies $x + y = 0$, and $1 - x - y = -1$ implies $x + y = 2$. Thus, the level curves are straight lines.

17.

    $c = (x + y)/(x - y)$ implies $cx - cy = x + y$, i.e., $(c - 1)x = (c + 1)y$ or $y = (c - 1)x/(c + 1)$. The general level curve is a line with slope $(c - 1)/(c + 1)$.

21.

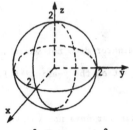

This is a sphere centered at the origin with radius 2 .

25. $c = x - y^2$ implies $y^2 = x - c$. The level curves are parabolas centered on the x-axis. (See below, left.) Its vertex is located at $(c,0)$ and the parabola opens up in the positive x-direction. Putting these curves together gives us a parabolic cylinder with the individual vertices located on the line $x = z$ in the xz-plane (below, right).

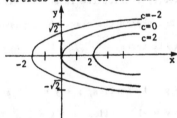

29.

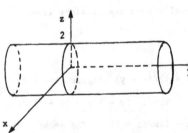

Since the variable y is missing, $z^2 + x^2 = 4$ represents a cylinder along the y-axis. On the xz-plane, $z^2 + x^2 = 4$ is a circle centered at the origin and with radius 2 .

33.

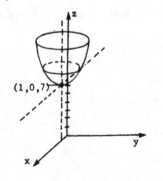

Completing the square yields $z = (x^2 - 2x + 1) + y^2 + 7$ or $z - 7 = (x - 1)^2 + y^2$, which is the same surface as in Exercise 31 except that this surface is shifted up by 7 units. When z is constant, the equation describes a circle centered at $(1,0)$ with radius $\sqrt{z - 7}$.

33. (continued)

Putting these level curves together gives us a paraboloid as in Example 7.

z   must be greater than or equal to   7 .

37.

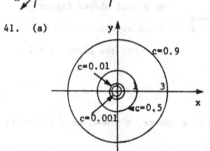

The y variable is missing from the equation, so the graph is a "cylinder". In the xz-plane, sketch the curve $z = \sin x$. Then move the curve along the y-axis in both directions.

41. (a)

$c = e^{-1/(x^2+y^2)}$ implies $\ln c = -1/(x^2 + y^2)$ or $-1/\ln c = x^2 + y^2$. This is a circle centered at the origin with radius $\sqrt{-1/\ln c}$ if $0 < c < 1$.

(b) An exponential of any number is positive, so the level curves of $c < 0$ do not exist. $x^2 + y^2$ is a positive quantity, so $-1/(x^2 + y^2) < 0$; therefore $e^{-1/(x^2+y^2)} < 1$, so no level curves exist for $c > 1$.

(c)

If $y = 0$, then $f(x,y) = z = e^{-1/x^2}$. $z'(x) = (-2/x^3)e^{-1/x^2}$, so there are no critical points. The graph is symmetric about the z-axis. The domain is $x \neq 0$. The exponent, $-1/x^2$, is always negative, so the range is $0 < z < 1$. $\lim_{x \to 0} z = 0$ and $\lim_{x \to \infty} z = 1$, so $z = 1$ is an asymptote.

41. (d)   In polar coordinates, the equation is $f(r,\theta) = e^{-1/r^2}$. This
equation is independent of θ , so the cross-section of any
vertical plane through the origin looks the same.

(e)

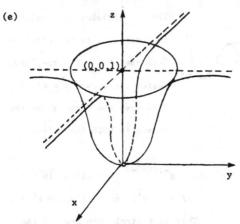

The surface consists of a
series of circles placed on
top of each other. The
radii of the circles in-
crease exponentially. Putting these together gives
us a bell-shaped figure
which asymptotically ap-
proaches the plane $z = 1$.

SECTION QUIZ

1.   Classify the following as the equation of a sphere, a plane, or a cylinder:

(a) $(x - 3)^2 + (y + 2)^2 + (z - 1)^2 = 4$

(b) $(x - 3) + (y + 2) + (z - 1) = 4$

(c) $2x - 5y + z = 0$

(d) $(x - 3)^2 + (y + 2) = 4$

(e) $y^2 - x^2 - 2 = 0$

2.   Sketch the surfaces in Question 1.

3.   (a)   Define a level curve.

(b)   Describe the level curves of $z = f(x,y)$ if the function's graph
is a plane.

(c)   Sketch the level curves for the function in Question 1(e) if
$c = -2, -1, 0, 1, 2$.

4.   Sketch the graph of $f(x,y) = x^3 - x$.

5.   Jealous Johnny saw Phil the Flirt dancing with his girl friend last night.
The revenge-minded Johnny dreamed up a devious plan. First, he was going
to trap Phil the Flirt inside a structure described by the equation
$(x - 1)^2 + (y + 1)^2 + (z - 2)^2 = 1$. Openings at $z = 2$ permitted
feathers to enter each hour to tickle Phil. Finally, if Phil should try
to escape, he would simply fall onto a smooth slide described by
$3x + y - z = 6$ and slide into a huge pool of anchovies.

(a) Describe the level curve for $f(x,y) = 2$ where the feathers entered.

(b) Sketch the structure which would entrap Phil the Flirt.

(c) Sketch the slide leading into the pool of anchovies.

ANSWERS TO PREREQUISITE QUIZ

1.   (a) $t \neq 3$

(b) $x \leqslant 1$ and $x \geqslant 2$

2.   (a)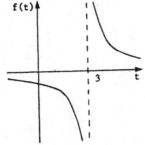

(b)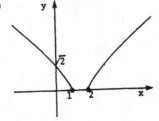

ANSWERS TO SECTION QUIZ

1.   (a) Sphere

     (b) Plane

     (c) Plane

     (d) Cylinder

     (e) Cylinder

2.   (a)

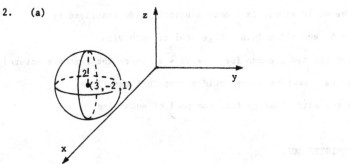

     (b)

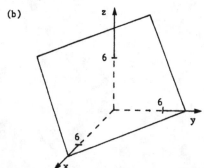

     (c)

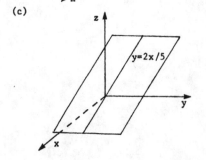

2.   (d)

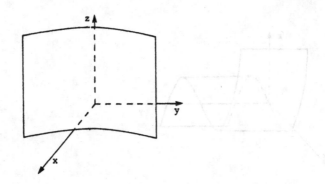

(e)

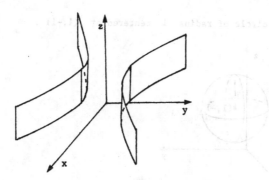

3.   (a) A level curve is the set of all (x,y) in the plane satisfying

f(x,y) = c , where c is a constant.

(b) The level curves are straight lines.

(c)

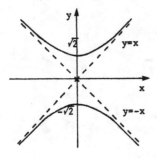

All of these level curves are the
same.

4.

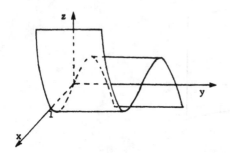

5.   (a) It is a circle of radius 1 centered at (1,−1) .

     (b)

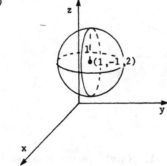

     (c)

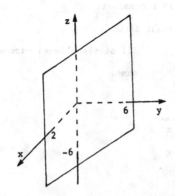

14.4 Quadric Surfaces

PREREQUISITES

1. Recall how to sketch level curves (Section 14.3).

2. Recall how to recognize equations of cylinders and sketch them
    (Section 14.3).

PREREQUISITE QUIZ

1. Let $z = x^2 + y^2$. Sketch the level curves in the xy-plane for $z = 0$,
    1 , and 3 .

2. Sketch the graph of $x^2 + 4y^2 = 4$ in space.

3. Describe how to recognize the equation of a cylinder.

GOALS

1. Be able to sketch quadric surfaces.

STUDY HINTS

1. Quadric surfaces. The surfaces in this section may all be sketched by
    using the method of sections, which was discussed in Section 14.4. Qua-
    dric surfaces may be classified by the form of the equation (see Study
    Hints 2,4,5,6,7, and 8). If the variables are interchanged, you should
    be able to analyze the equations in a manner analogous to the text's
    discussion. The difference is that these surfaces will be "lying on
    their sides" rather than "standing up." For example, $z = 4x^2 - 3y^2$
    and $y = 4z^2 - 3x^2$ are both hyperbolic paraboloids.

2. Hyperbolic paraboloid. The general equation is $z = ax^2 - by^2$, where
    a and b have the same sign. The level curves for constant z are
    hyperbolas, except when $z = 0$. The hyperbolas are centered about a
    different axis when z changes sign. The hyperbolas degenerate to two
    straight lines when $z = 0$. See Example 1 and Exercise 17.

3. Saddle points. At these points, the surfaces are rising in certain
    directions and falling in others. For example, in Fig. 14.4.2, the
    function has a local minimum if we set $y = 0$. If $x = 0$, we have a
    local maximum. More details are discussed in Chapter 16.

4. Hyperboloid of two sheets. The general equation is $x^2/a^2 + y^2/b^2 -
    z^2/c^2 = -1$. In each plane parallel to the xy-plane, the level curve
    is an ellipse with equation $x^2/a^2 + y^2/b^2 = -1 + z^2/c^2$. Thus, no
    level curve exists if $z^2/c^2 < 1$. It is easier to figure out this
    fact rather than memorizing it. For example, at $z = 0$, there are no
    points, so there must be a "gap." Note that in planes parallel to the
    xz- and yz-planes, the level curve is a hyperbola. See Example 4.

5. Ellipsoid. The general equation is $x^2/a^2 + y^2/b^2 + z^2/c^2 = 1$. These
    surfaces are "footballs" with intercepts at $x = \pm a$, $y = \pm b$, $z = \pm c$.
    Any level curve is an ellipse which gets smaller as you move away from
    the planes defined by the coordinate axes, i.e., the xy-, xz-, or yz-
    planes, to the intercepts. See Example 5.

6. Hyperboloid of one sheet. The general equation is $x^2/a^2 + y^2/b^2 -
    z^2/c^2 = 1$. Note the similarities and differences with the two-sheeted
    hyperboloid. In the general form, one equation's right-hand side equals
    1 and the other has -1 . Parallel to the xy-plane, both are ellipses
    for constant z . However, ellipses exist for all z in the one-sheet
    form. See Example 6.

7. <u>Elliptic paraboloid</u>. The general form is $z = ax^2 + by^2$, where a
    and b have the same sign. If z is constant, the level curves are
    ellipses. The level curves are smallest when $z = 0$ (a point) and
    the level curves get larger with $|z|$. Level curves in planes parallel
    to the xz- and yz-planes are parabolas. These surfaces are bowl shaped.
    See Exercise 18.

8. <u>Elliptic cone</u>. The general form is $x^2/a^2 + y^2/b^2 - z^2/c^2 = 0$. By
    rearranging, we get $z^2 = c^2x^2/a^2 + c^2y^2/b^2$. As with the elliptic
    paraboloid, the level curves for constant z are again ellipses. The
    general shape, as the name suggests, is a cone. See Exercise 25.

9. <u>Origin of conics</u>. Examples 8 and 9 prove that slices of a cone are
    conics. Except in honors classes, you will probably not have to
    reproduce similar arguments.

SOLUTIONS TO EVERY OTHER ODD EXERCISE

1.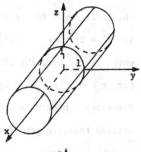

$y^2 + z^2 = 1$ represents a unit
circular cylinder along the x-axis.

5.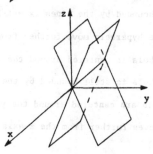

$z^2 - 8y^2 = 0$ if $z = \pm\sqrt{8}y$. Thus,
in three dimensions, this represents
two intersecting planes. The planes
intersect on the x-axis.

9.

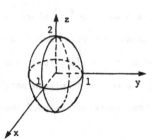

$x^2 + y^2 + z^2/2 = 1$ is an ellipsoid (see Example 5). The intercepts are $x = \pm 1$, $y = \pm 1$, and $z = \pm 2$.

13.

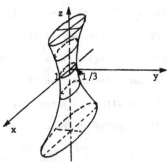

Rearrangement yields $x^2 + 9y^2 = z^2 + 1$ which is $x^2 + 9y^2 = 1$ in the xy-plane. This is an ellipse with x-intercepts, ± 1, and y-intercepts, $\pm 1/3$. Note that the graph is symmetric to the xy-plane. As z approaches $\pm \infty$, the cross-sections are ellipses of increasing size. The surface is a hyperboloid of one sheet. (See Example 6.)

17. (a)

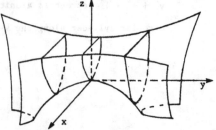

The level curve for $c = 0$ is $x^2 = 2y^2$ or $x = \pm\sqrt{2}y$, which is two straight lines. For $c > 0$, $c = x^2 - 2y^2$ describes a hyperbola. Rewriting this, we get $1 = x^2/c - 2y^2/c$, so the hyperbola is bounded by the lines $x = \pm\sqrt{2}y$. As c increases, the vertices of the hyperbola move further from the z-axis. For $c > 0$, the hyperbola is centered around the x-axis. If $c < 0$, then the hyperbola is still bounded by the lines $x = \pm\sqrt{2}y$, but these hyperbola are centered around the y-axis. For decreasing c, the hyperbola moves further from the z-axis.

17. (b)

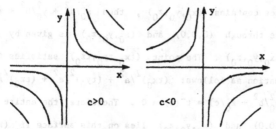

c>0   c<0

When c = 0 , the level curve is x = 0 and y = 0 , two straight

lines. For c > 0 , the hyperbola exists in the first and third

quadrants. For c < 0 , the hyperbola lies in the second and

fourth quadrants. For c = 2 , 1 , 0 , -1 , -2 , the level curves

are the same as those in Fig. 14.4.1, except that the curves are

rotated $45°$ to the left. Thus, the surface is also rotated $45°$,

so z = xy determines a hyperbolic paraboloid.

21.

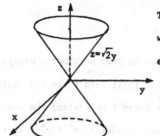

$z = \sqrt{2}y$

This cone is composed of cross-sections

which are ellipses. For constant z , the

ellipse is $1 = x^2/z^2 + y^2/(z^2/2)$.

25. (a) The equation can be rewritten as $x^2/a^2 + y^2/b^2 = z^2/c^2$ or

$c^2x^2/a^2 + c^2y^2/b^2 = z^2$. For a constant z , we have $c^2x^2/a^2z^2 +$

$c^2y^2/b^2z^2 = 1$, which is an ellipse centered around the z-axis.

The axes of the ellipse have lengths 2az/c and 2bz/c .

(b) When x = 0 , the equation is $y^2/b^2 = z^2/c^2$ or y = ±bz/c .

When y = 0 , the equation is $x^2/a^2 = z^2/c^2$ or x = ±az/c . In

both cases, the cross-sections are two straight lines.

25. (c) If the surface contains (x_0, y_0, z_0) , then $x_0^2/a^2 + y_0^2/b^2 + z_0^2/c =$

0 . The line through $(0,0,0)$ and (x_0, y_0, z_0) is given by

$(x,y,z) = t(x_0, y_0, z_0)$. The point (tx_0, ty_0, tz_0) satisfies the

original equation as follows: $(tx_0)^2/a^2 + (ty_0)^2/a^2 + (tz_0)^2/c^2 =$

$t^2(x_0^2/a^2 + y_0^2/b^2 + z_0^2/c) = t^2(0) = 0$. Therefore, the entire line

through $(0,0,0)$ and (x_0, y_0, z_0) lies on this surface if (x_0, y_0, z_0)

is on it.

SECTION QUIZ

1. Sketch the following surfaces:

(a) $x^2/4 + y^2/9 + z^2/16 = 1$

(b) $9x^2 + 9y^2 - z^2 = 4$

(c) $9x^2 + 9y^2 - z^2 = -4$

(d) $3x^2 + 2y^2 - z = 0$

(e) $4x^2 + y^2/9 = z^2/4$

2. An alien spacecraft has just crash-landed in your neighbor's backyard.
These feather-covered aliens wish to have their ship's cracked shell
repaired. Unfortunately, your neighbor doesn't understand the cryptic
message given to him. The message reads: " σΗελλ: $x^2 + y^2 + z^2 = 1$
if $z \leqslant 0$; $4x^2 + 4y^2 + z^2 = 4$ if $z > 0$. " Help your neighbor sketch
the shell so that he can help the chicken-like aliens repair their ship.

ANSWERS TO PREREQUISITE QUIZ

1.

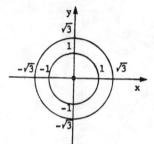

2.

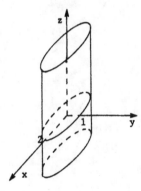

3.   One of the variables, x , y , or z , is missing from the equation.

ANSWERS TO SECTION QUIZ

1.   (a)

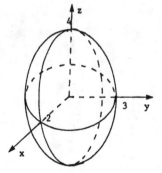

1. (b)

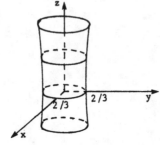

(c)

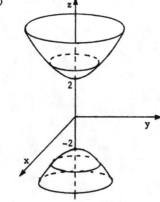

(d)

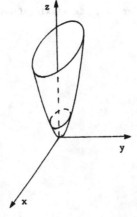

1.   (e)

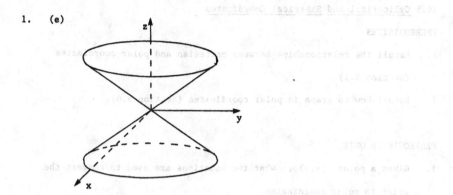

2.

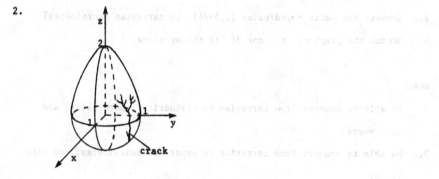

14.5 Cylindrical and Spherical Coordinates

PREREQUISITES

1.   Recall the relationships between cartesian and polar coordinates
      (Section 5.1).

2.   Recall how to graph in polar coordinates (Section 5.6).

PREREQUISITE QUIZ

1.   Given a point (x,y) , what two equations are used to convert the
      point to polar coordinates.

2.   Convert the polar coordinates $(2,3\pi/4)$ to cartesian coordinates?

3.   Sketch the graph of $r = \cos 3\theta$ in the xy-plane.

GOALS

1.   Be able to convert from cartesian to cylindrical coordinates, and
      vice versa.

2.   Be able to convert from cartesian to spherical coordinates, and vice
      versa.

STUDY HINTS

1.   Cylindrical coordinates. You should know how to convert back and forth
      between cylindrical and cartesian coordinates. The formulas to convert
      from cylindrical coordinates are $x = r \cos \theta$, $y = r \sin \theta$, $z = z$.
      To convert from cartesian coordinates, use $r = \sqrt{x^2 + y^2}$ and $\theta =$
      $\tan^{-1}(y/x)$ if $x \geqslant 0$, $\theta = \tan^{-1}(y/x) + \pi$ if $x < 0$. Plotting the
      point in xy-coordinates will help you decide what θ should be.

2. <u>Spherical coordinates</u>. Know how to convert back and forth between spherical and cartesian coordinates. In addition, you should know the geometry associated with the variables ρ , ϕ , and θ : ρ is the distance from the origin; ϕ is the angle from the positive z-axis; θ is the same angle as in cylindrical coordinates. The formulas you need to know are $x = \rho \sin \phi \cos \theta$, $y = \rho \sin \phi \sin \theta$, $z = \rho \cos \phi$, $\rho = \sqrt{x^2 + y^2 + z^2}$, and $\phi = \cos^{-1}(z/\sqrt{x^2 + y^2 + z^2})$. θ is the same as in cylindrical coordinates. Again, plotting the point in xyz-coordinates will help you decide what θ should be.

3. <u>Graphs of r = constant</u>. Note that r = constant in cylindrical coordinates describes a cylinder and that r = constant in spherical coordinates describes a shpere. You may have suspected this from the name of the coordinate system.

SOLUTIONS TO EVERY OTHER ODD EXERCISE

1.

Since $x \geqslant 0$, we use $\theta = \tan^{-1}(y/x)$ rather than $\theta = \tan^{-1}(y/x) + \pi$. Also, $r = \sqrt{x^2 + y^2}$ and z = z . Thus, $r = \sqrt{1^2 + (-1)^2} = \sqrt{2}$ and $\theta = \tan^{-1}(-1) = -\pi/4$. Therefore, (1,-1,0) converts to $(\sqrt{2}, -\pi/4, 0)$.

5.

Using the same formulas as in Exercise 1, we get $r = \sqrt{6^2 + 0^2} = 6$ and $\theta = \tan^{-1}(0) = 0$. Therefore, (6,0,-2) converts to (6,0,-2) .

9.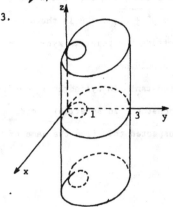

Use the formulas $x = r \cos \theta$, $y = r \sin \theta$, and $z = z$. In this case, $x = (-1) \cos(\pi/6) = -\sqrt{3}/2$ and $y = (-1) \sin(\pi/6) = -1/2$. Thus, $(-1,\pi/6,4)$ converts to $(-\sqrt{3}/2,-1/2,4)$.

13.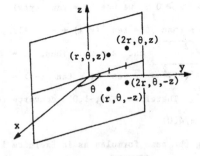

Since z is missing, $r = 1 + 2 \cos \theta$ represents a cylinder in the z-axis direction. Note that $r = 1 + 2 \cos \theta$ in polar coordinates looks like a cardioid.

17.

The effect of replacing z by $-z$ is to reflect the point across the xy-plane. The effect of replacing r by $2r$ is to move the point twice as far from the z-axis. Replacing (r,θ,z) by $(2r,\theta,-z)$ moves the points twice as far from the z-axis and reflects it across the xy-plane.

21.

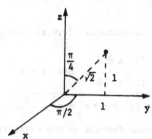

$\rho = \sqrt{x^2 + y^2 + z^2} = \sqrt{0 + 1 + 1} = \sqrt{2}$;

$\phi = \cos^{-1}(z/\rho) = \cos^{-1}(1/\sqrt{2}) = \pi/4$. θ

is the rotation in the xy-plane from the

positive x-axis. In this case, $\theta =$

$\tan^{-1}(1/0) = \tan^{-1}(\pm\infty) = \pi/2$. Thus,

(0,1,1) converts to $(\sqrt{2}, \pi/2, \pi/4)$.

25.

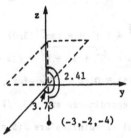

Since $x < 0$, we use $\theta = \tan^{-1}(y/x) +$

π , $\rho = \sqrt{x^2 + y^2 + z^2}$, and $\phi =$

$\cos^{-1}(z/\rho)$. Thus, $\rho = \sqrt{9 + 4 + 16} =$

$\sqrt{29}$, $\theta = \tan^{-1}(-2/-3) + \pi \approx 3.73$, and

$\phi = \cos^{-1}(-4/\sqrt{29}) \approx 2.41$. Thus,

(-3,-2,-4) converts to $(\sqrt{29}, 3.73, 2.41)$.

29.

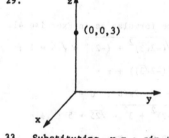

Use $x = \rho \sin \phi \cos \theta$, $y =$

$\rho \sin \phi \sin \theta$, and $z = \rho \cos \phi$. Thus,

$x = 3 \sin (0) \cos (2\pi) = 0$, $y =$

$3 \sin (0) \sin (2\pi) = 0$, and $z =$

$3 \cos (0) = 3$. Therefore, $(3, \pi/2, 0)$

converts to (0,0,3) .

33. Substituting $x = \rho \sin \phi \cos \theta$ and $z = \rho \cos \phi$ converts $xz = 1$ to

$\rho^2 \sin \phi \cos \phi \cos \theta = 1$ in spherical coordinates. That is,

$\rho^2 \sin 2\phi \cos \theta = 2$ or $\rho^2 = 2/\sin 2\phi \cos \theta$. (We used the identity

$\sin 2\phi = 2 \sin \phi \cos \phi$.)

37. Replacing (ρ, θ, ϕ) by $(2\rho, \theta, \phi)$ extends a point twice as far from the

origin along the same line through the origin because θ and ϕ

do not change.

41. We are given (x,y,z) . The cylindrical coordinates (r,θ,z) are given

by $r = \sqrt{x^2 + y^2}$, $\theta = \begin{cases} \tan^{-1}(y/z) & \text{if } x > 0 \\ \tan^{-1}(y/x) + \pi & \text{if } x < 0 \end{cases}$, and $z = z$. The

spherical coordinates (ρ,θ,ϕ) are given by $\rho = \sqrt{x^2 + y^2 + z^2}$,

$\phi = \cos^{-1}(z/\rho)$ and θ is given by the same formula as the cylindrical

coordinate.

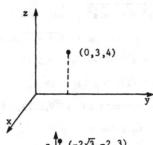

$r = \sqrt{0^2 + 3^2} = 3$; $\theta = \tan^{-1}(3/0)$ implies

$\theta = \pi/2$; $\rho = \sqrt{0^2 + 3^2 + 4^2} = \sqrt{25} = 5$;

$\phi = \cos^{-1}(4/5) \approx 0.64$. Therefore, the

cylindrical coordinates are $(3,\pi/2,4)$ and

the spherical coordinates are $(5,\pi/2,0.64)$.

45.   Using the same formulas as in Exercise 41,

we get $r = \sqrt{(-2\sqrt{3})^2 + (-2)^2} = \sqrt{16} = 4$;

$\theta = \tan^{-1}(-2/(-2\sqrt{3})) + \pi =$

$\tan^{-1}(1/\sqrt{3}) + \pi = 7\pi/6$; $\rho =$

$\sqrt{(-2\sqrt{3})^2 + (-2)^2 + 3^2} = \sqrt{25} = 5$; $\phi =$

$\cos^{-1}(3/5) \approx 0.93$. Therefore, the

cylindrical coordinates are $(4,7\pi/6,3)$ and the spherical coordinates

are $(5,7\pi/6,0.93)$.

49. Given (r,θ,z) , the conversion formulas for cartesian coordinates are

$x = r\cos\theta$, $y = r\sin\theta$, and $z = z$. The conversion formulas for

spherical coordinates are $\rho = \sqrt{r^2 + z^2}$, $\theta = \theta$, and $\phi = \cos^{-1}(z/\rho)$.

Therefore, we have $x = (0)(\cos(\pi/4)) = 0$; $y = (0)(\sin(\pi/4)) = 0$; $\rho =$

$\sqrt{0^2 + 1^2} = 1$; $\phi = \cos^{-1}(1/1) = 0$. Thus, the cartesian coordinates are

$(0,0,1)$ and the spherical coordinates are $(1,\pi/4,0)$.

53. We are given the spherical coordinates (ρ,θ,ϕ) . The cartesian coordinates are given by $x = \rho \sin \phi \cos \theta$, $y = \rho \sin \phi \sin \theta$, and $z = \rho \cos \phi$. After finding the cartesian coordinates, the cylindrical coordinates are given by $r = \sqrt{x^2 + y^2}$, $\theta = \begin{cases} \tan^{-1}(y/x) & \text{if } x \geq 0 \\ \tan^{-1}(y/x) + \pi & \text{if } x < 0 \end{cases}$,

$z = z$.

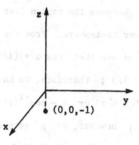

$x = (1) \sin(\pi) \cos(\pi/2) = 0$;

$y = (1) \sin(\pi) \sin(\pi/2) = 0$;

$z = (1) \cos(\pi) = -1$, so the cartesian coordinates are $(0,0,-1)$. $r = \sqrt{0^2 + 0^2} = 0$, and since there is no rotation in the xy-plane, θ can have any value, so the cylindrical coordinates are $(0,\theta,-1)$.

57. Using the same formulas as in Exercise 53, we get $x = (-1) \sin(\pi/6) \cos(\pi) = (-1)(1/2)(-1) = 1/2$; $y = (-1)\sin(\pi/6) \times \sin(\pi) = 0$; $z = (-1) \cos(\pi/6) = -\sqrt{3}/2$; $r = \sqrt{(1/2)^2 + 0^2} = 1/2$; $\theta = \tan^{-1}(0/(1/2)) = 0$. Therefore, the cartesian coordinates are $(1/2,0,-\sqrt{3}/2)$ and the cylindrical coordinates are $(1/2,0,-\sqrt{3}/2)$.

61. (a) The length of $x\underline{i} + y\underline{j} + z\underline{k}$ is $\sqrt{x^2 + y^2 + z^2}$. The formula for the vector's length is precisely the definition of the spherical coordinate ρ .

(b) By the definition of the dot product, $\underline{v} \cdot \underline{k} = x(0) + y(0) + z(1) = z$. Also, $\|\underline{v}\| = \sqrt{x^2 + y^2 + z^2}$, so $\cos^{-1}(\underline{v} \cdot \underline{k}/\|\underline{v}\|) = \cos^{-1}(z/\sqrt{x^2 + y^2 + z^2})$, which is the definition of the spherical coordinate ϕ .

61. (c)

$\underline{u} \cdot \underline{i} = x(1) + y(0) + 0(0) = x$, and $\|\underline{u}\| = \sqrt{x^2 + y^2 + 0^2}$, so $\cos^{-1}(\underline{u} \cdot \underline{i}/\|\underline{u}\|) = \cos^{-1}(x/\sqrt{x^2 + y^2})$. From the diagram, we see that $\cos^{-1}(x/\sqrt{x^2 + y^2}) = \theta = \tan^{-1}(y/x)$, which is the definition of θ given in the text.

65.

Note that ϕ will be between $\pi/2$ and π because the region lies in the lower hemisphere. From the triangle, we see that $\cos \alpha = (d/6)/(d/2) = 1/3$; therefore, we have $\pi - \alpha \leq \phi \leq \pi$ or $\pi - \cos^{-1}(1/3) \leq \phi \leq \pi$. Now, ρ can be as large as $d/2$; however, as ρ gets smaller, its lower limit depends on ϕ . Pick any ϕ , then $\phi + \beta = \pi$ and according to the diagram $\cos \beta = (d/6)/\rho$. Rearrangement gives $d/6 \cos \beta = \rho = d/6 \cos(\pi - \phi) = -d/6 \cos \phi$. Therefore, $-d/6 \cos \phi \leq \rho \leq d/2$. So far, we have described the cross-section in one quadrant. The entire volume requires a revolution around the z-axis, so its description is $-d/6 \cos \phi \leq \rho \leq d/2$, $0 \leq \theta \leq 2\pi$, and $\pi - \cos^{-1}(1/3) \leq \phi \leq \pi$.

SECTION QUIZ

1. Explain the geometric meaning of the spherical coordinates ρ , θ , and ϕ .

2. The cartesian coordinates $(-1,1,-5/2)$ do not convert to the cylindrical coordinates $(\sqrt{2},-\pi/4,-5/2)$ even though $\tan^{-1}(-1) = -\pi/4$. Why?

3. If a point lies on the xy-plane, then ϕ is: (choose one)

    (a) unknown because we don't know x and y .

    (b) $\pi/2$.

    (c) 90 .

4. An inexperienced sailor has become ship-wrecked on a Greek island and
    has been unable to radio for help. However, upon exploring the island,
    the sailor comes upon an odd-looking capped bottle. As he uncaps the
    bottle, he is surprised by an emerging genie. Her last master was Euclid,
    and she only understands how to locate places with cylindrical coordinates.
    The sailor knows from his latitude and longitude charts that the
    spherical coordinates of New York City are $(1,-72^{\circ},49^{\circ})$. (Distances
    are measured in Earth-radius units.)

    (a) What are the corresponding cylindrical coordinates?

    (b) What are the corresponding cartesian coordinates?

ANSWERS TO PREREQUISITE QUIZ

1. $r = \sqrt{x^2 + y^2}$, and $\theta = \tan^{-1}(y/x)$ if $x \geq 0$, $\theta = \tan^{-1}(y/x) + \pi$
    if $x < 0$.

2. $(-\sqrt{2},\sqrt{2})$

3.

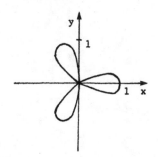

ANSWERS TO SECTION QUIZ

1. ρ is the distance from the origin; θ is the angle in the xy-plane, measured counterclockwise from the x-axis; ϕ is the angle measured from the positive z-axis.

2. θ should be $3\pi/4$.

3. b

4. (a) (0.75,-1.26,0.66)

    (b) (0.23,-0.71,0.66)

14.6 <u>Curves in Space</u>

PREREQUISITES

1. Recall how to sketch curves in the plane described by parametric
    equations (Sections 2.4 and 10.4).

2. Recall the basic rules for differentiation (Chapters 1 and 2).

PREREQUISITE QUIZ

1. Sketch the curves described by the following sets of parametric
    equations:

    (a) $x = \cos t$, $y = \sin t$, $0 \leq t \leq 2\pi$

    (b) $y = 2t$, $x = 3t - 1$

2. Differentiate the following functions:

    (a) $(x^3 + x)(x^2 + 1)$

    (b) $x/(x - 1)$

    (c) $\sin (3t - 1)$

GOALS

1. Be able to sketch curves in space.

2. Be able to differentiate vector functions.

STUDY HINTS

1. <u>Parametric lines</u>. One of the simplest curves to write parametrically
    is the straight line. As shown by Example 2, the parameter occurs with
    only one power. In (a), t occurs only to the first power. In (b),
    t occurs in the third power.

2. <u>Sketching curves</u>. As with surfaces, we sketch curves with a technique
    similar to the method of sections. In many cases, the curve is drawn
    in the xy-plane and then lifted to the appropriate height.

3. <u>Terminology</u>. The object $(f(t), g(t), h(t))$ may be called a parametric curve, a vector function, or a triple of scalars.

4. <u>Differentiating vector functions</u>. The derivative of $\underline{\sigma}(t) = f(t)\underline{i} + g(t)\underline{j} + h(t)\underline{k}$ is $\underline{\sigma}'(t) = f'(t)\underline{i} + g'(t)\underline{j} + h'(t)\underline{k}$. This formula should be memorized.

5. <u>Differentiation rules</u>. In the box on p. 740, note that there are three product rules for vectors — scalar multiplication, dot product, and cross product. These rules are very similar to the rules you learned in Chapters 1 and 2.

6. <u>Physical application</u>. Suppose that $(f(t), g(t), h(t))$ describes a parametric curve. The derivative $(f'(t), g'(t), h'(t))$ is known as a velocity vector and the second derivative is the acceleration vector. The length of the velocity vector is called the speed. This is analogous to the material presented in Section 10.4.

SOLUTIONS TO EVERY OTHER ODD EXERCISE

1.

Rearrangement gives $y/4 = \cos t$; therefore, $x^2 + (y/4)^2 = \sin^2 t + \cos^2 t = 1$. This is an ellipse with intercepts $(\pm 1, 0)$ and $(0, \pm 4)$. The curve is traced out clockwise starting at $(0, 4)$.

5.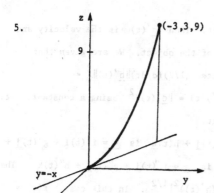

Note that $x = -t$, $y = t$ represents the straight line $y = -x$ in the xy-plane. Thus, $(x,y,z) = (-t,t,t^2)$ moves along a curve over this line with the z component moving like t^2.

9.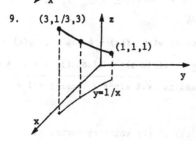

In the xy-plane, we have the hyperbola $xy = 1$. Move its points t units in the z-direction to get the desired curve.

13. If $\underline{\sigma}(t) = f(t)\underline{i} + g(t)\underline{j} + h(t)\underline{k}$, then $\underline{\sigma}'(t) = f'(t)\underline{i} + g'(t)\underline{j} + h'(t)\underline{k}$. Since $\underline{\sigma}(t) = 3(\cos t)\underline{i} - 8(\sin t)\underline{j} + e^t\underline{k}$, we get $\underline{\sigma}'(t) = -3(\sin t)\underline{i} - 8(\cos t)\underline{j} + e^t\underline{k}$. Similarly, $\underline{\sigma}''(t) = -3(\cos t)\underline{i} + 8(\sin t)\underline{j} + e^t\underline{k}$.

17. By the cross-product rules for differentiation, $(d/dt)[\underline{\sigma}_1(t) \times \underline{\sigma}_2(t)] = \underline{\sigma}_1'(t) \times \underline{\sigma}_2(t) + \underline{\sigma}_1(t) \times \underline{\sigma}_2'(t) = [t^3\cos t(-2 + \csc^2 t) - 3t^2\sin t (2 + \csc^2 t)]\underline{i} + [t^2 e^{-t}(3 - t) + 2t^2 e^t(t + 3)]\underline{j} + [e^t \csc t(1 - \cot t) - e^{-t}(\cos t + \sin t)]\underline{k}$. On the other hand, $\underline{\sigma}_1(t) \times \underline{\sigma}_2(t) = (-2t^3\sin t - t^3\csc t)\underline{i} + (t^3 e^{-t} + 2t^3 e^t)\underline{j} + (e^t\csc t - e^{-t}\sin t)\underline{k}$. Thus, $(d/dt)[\underline{\sigma}_1(t) \times \underline{\sigma}_2(t)] = [(d/dt)(-2t^3\sin t - t^3\csc t)]\underline{i} + [(d/dt)(t^3 e^{-t} + 2t^3 e^t)]\underline{j} + [(d/dt)(e^t\csc t - e^{-t}\sin t)]\underline{k}$. Completing the differentiation yields the same result as before.

21. Let $\underline{\sigma}(t)$ be the path of an object. Then $\underline{\sigma}'(t)$ is the velocity and $\underline{\sigma}''(t)$ is the acceleration vector of the object. We are given that $\underline{\sigma}'(t) \cdot \underline{\sigma}''(t) = 0$ for all t. Since $(1/2)(d/dt)\|\underline{\sigma}'(t)\|^2 = \underline{\sigma}'(t) \cdot \underline{\sigma}''(t) = 0$, we have $(\underline{\sigma}' \cdot \underline{\sigma}')(t) = \|\underline{\sigma}'(t)\|^2$ being a constant, i.e., the speed of the object is constant.

25. The velocity vector of $f(t)\underline{i} + g(t)\underline{j} + h(t)\underline{k}$ is $\underline{v} = f'(t)\underline{i} + g'(t)\underline{j} + h'(t)\underline{k}$. The acceleration vector is $\underline{a} = f''(t)\underline{i} + g''(t)\underline{j} + h''(t)\underline{k}$. The speed is $[(f'(t))^2 + (g'(t))^2 + (h'(t))^2]^{1/2}$. In this case, $f(t) = 2t - 1$, $g(t) = t + 2$, and $h(t) = t$. Thus, $\underline{v} = 2\underline{i} + \underline{j} + \underline{k}$, $\underline{a} = \underline{0}$, and the speed is $\sqrt{4 + 1 + 1} = \sqrt{6}$.

29. Using the same formulas as in Exercise 25 with $f(t) = 4\cos t$, $g(t) = 2\sin t$, and $h(t) = t$, we get $\underline{v} = -4(\sin t)\underline{i} + 2(\cos t)\underline{j} + \underline{k}$, $\underline{a} = -4(\cos t)\underline{i} - 2(\sin t)\underline{j}$, and the speed is $\sqrt{16\sin^2 t + 4\cos^2 t + 1} = \sqrt{5 + 12\sin^2 t}$.

33. For a parametric curve $(f(t),g(t),h(t))$, the velocity vector is $(f'(t),g'(t),h'(t))$ and the acceleration vector is $(f''(t),g''(t),h''(t))$. The tangent line is given by $(f(t_0),g(t_0),h(t_0)) + (t - t_0)(f'(t_0), g'(t_0),h'(t_0))$. In this case, $\underline{v}(t) = (6,6t,3t^2)$; $\underline{a}(t) = (0,6,6t)$; therefore, the tangent line is $(0,0,0) + (6,0,0)t = (6,0,0)t$.

37. Using the same method as in Exercise 33, we get $\underline{v}(t) = (\sqrt{2},e^t,-e^{-t})$; $\underline{a}(t) = (0,e^t,e^{-t})$; therefore, the tangent line is $(0,1,1) + (\sqrt{2},1,-1)t$.

41. We integrate $\underline{\sigma}'(t) = (A,B,C)$ to get $\underline{\sigma}(t) = (At + c_1, Bt + c_2, Ct + c_3)$. Substituting $t = 0$ gives $\underline{\sigma}(0) = (c_1,c_2,c_3)$. Thus, we get the following curves:

41. (continued)

   (a) $\underline{\sigma}(t) = (1,0,1)t$.

   (b) $\underline{\sigma}(t) = (1,2,3) + (-1,1,1)t$.

   (c) $\underline{\sigma}(t) = (-1,1,1)t$.

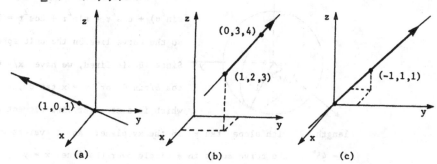

(a)   (b)   (c)

45. (a) Since $\underline{\sigma}_1(t)$ and $\underline{\sigma}_2(t)$ satisfies the differential equation, then
   it must be true that $\underline{\sigma}_1''(t) = -\underline{\sigma}_1(t)$ and $\underline{\sigma}_2''(t) = -\underline{\sigma}_2(t)$. Now,
   let $\underline{\sigma}(t) = A_1\underline{\sigma}_1(t) + A_2\underline{\sigma}_2(t)$. We have $\underline{\sigma}''(t) = A_1\underline{\sigma}_1''(t) + A_2\underline{\sigma}_2''(t) =$
   $-A_1\underline{\sigma}_1(t) - A_2\underline{\sigma}_2(t) = -\underline{\sigma}(t)$.

   (b) Section 8.1 provides the solution to the spring equation $d^2x/dt^2 =$
   $-\omega^2 x$ as $x = A \cos \omega t + B \sin \omega t$, for constants A and B .
   Here, $\omega = 1$, and each component of $\underline{\sigma}(t)$ must satisfy the spring
   equation, so the most general form of $\underline{\sigma}(t)$ is $(A_1 \cos t +$
   $B_1 \sin t, A_2 \cos t + B_2 \sin t, A_3 \cos t + B_3 \sin t)$. A_1 , A_2 , A_3 ,
   B_1 , B_2 , and B_3 are all constants.

49. (a)

$(x^2 + y^2) + z^2 = \sin^2\phi(\cos^2 t + \sin^2 t) +$
$\cos^2\phi = \sin^2\phi + \cos^2\phi = 1$, so the
curve lies on the unit sphere. Since ϕ
is fixed, the curve lies parallel to
the xy-plane; therefore, the curve is a
circle. The center is at $(0,0,\cos\phi)$

49. (a) (continued)

and the radius is $\sin \phi$. For $\phi = 45°$, the circle lies in

the plane $z = \sqrt{2}/2$.

(b)

$(x^2 + y^2) + z^2 = \sin^2 t(\cos^2 \theta +$

$\sin^2 \theta) + \cos^2 t = \sin^2 t + \cos^2 t = 1$,

so the curve lies on the unit sphere.

Since θ is fixed, we have $x/y =$

$\cos \theta / \sin \theta$ or $y = x \tan \theta$,

which is a straight line segment of

length 2 with slope $\tan \theta$ in the xy-plane. As t varies with

$\theta = 45°$, the curve moves in a circle over the line $x = y$

beginning from $(0,0,1)$.

53. Let $\underline{\sigma}_1(t) = f_1(t)\underline{i} + g_1(t)\underline{j} + h_1(t)\underline{k}$ and $\underline{\sigma}_2(t) = f_2(t)\underline{i} + g_2(t)\underline{j} +$

$h_2(t)\underline{k}$. Then $\underline{\sigma}_1(t) + \underline{\sigma}_2(t) = [f_1(t) + f_2(t)]\underline{i} + [g_1(t) + g_2(t)]\underline{j} +$

$[h_1(t) + h_2(t)]\underline{k}$, and $(d/dt)[\underline{\sigma}_1(t) + \underline{\sigma}_2(t)] = (d/dt)[f_1(t) + f_2(t)]\underline{i} +$

$(d/dt)[g_1(t) + g_2(t)]\underline{j} + (d/dt)[h_1(t) + h_2(t)]\underline{k} = [f_1'(t) + f_2'(t)]\underline{i} +$

$[g_1'(t) + g_2'(t)]\underline{j} + [h_1'(t) + h_2'(t)]\underline{k}$ by the sum rule for real-valued

functions. Rearranging the terms, we get $(d/dt)[\underline{\sigma}_1(t) + \underline{\sigma}_2(t)] =$

$\underline{\sigma}_1'(t) + \underline{\sigma}_2'(t)$.

57. (a) $\underline{\sigma}(t) \cdot \underline{u} = 0$ implies $\|\underline{\sigma}(t)\|\|\underline{u}\| \cos \theta = 0$ so $\theta = \pm\pi/2$ (provided

$\underline{\sigma}(t) \neq \underline{0}$) . Therefore, $\underline{\sigma}(t)$ is always perpendicular to $\underline{u}$, and

$\underline{\sigma}(t)$ describes any curve in a plane through the origin and

perpendicular to $\underline{u}$.

(b) By differentiating, we get $\underline{\sigma}'(t) \cdot \underline{u} + \underline{\sigma}(t) \cdot \underline{u}' = 0$. Since $\underline{u}$ is

constant, $\underline{u}' = 0$, and $\underline{\sigma}'(t) \cdot \underline{u} = 0$. Thus, $\underline{\sigma}'(t)$ is always per-

pendicular to $\underline{u}$. Again, this implies that $\underline{\sigma}(t)$ is any curve in

a plane perpendicular to $\underline{u}$. This plane need not go through the

origin.

57. (c) $\underline{\sigma}(t) \cdot \underline{u} = \|\underline{\sigma}(t)\| \cdot \|\underline{u}\| \cos\theta = b\|\underline{\sigma}(t)\|$. Since $\underline{u}$ is a unit vector

the equation reduces to $\cos\theta = b$ or $\theta = \cos^{-1}b$, where θ is

between $-\pi/2$ and $\pi/2$. Therefore, $\underline{\sigma}(t)$ is any curve lying in

the cone with $\underline{u}$ as its axis and vertex angle $2\cos^{-1}b$.

SECTION QUIZ

1. Sketch the curve $\underline{\sigma}(t) = (t\sin t, t\sin t, t)$.

2. Find $\underline{\sigma}'(t)$ and $\underline{\sigma}''(t)$ for the curve in Question 1.

3. Sketch the curve $\underline{\sigma}(t) = (\sin 2t + 3, \cos 2t + 1, t)$. [Hint: How does
   this compare to the curve $(\sin 2t, \cos 2t, t)$?]

4. Let $\underline{u}(t) = 3t\underline{i} + e^t\underline{j} + (\cos t)\underline{k}$ and $\underline{v}(t) = (\ln t)\underline{i} - (\sin^{-1}t)\underline{j} +$
   $(t/(1 - t))\underline{k}$. Compute:

   (a) $(d/dt)(\underline{u} \cdot \underline{v})$

   (b) $(d/dt)(\underline{u} \times \underline{v})$

   (c) $(d/dt)\underline{u}(t^2)$

   (d) $(d/dy)[\underline{u}(y) - \underline{v}(y)]$

5. Cute Karen, the fairest lady of the land, had been imprisoned in a tall tower
   by a mean magician. Along comes Prince Charming on his white stallion to the
   rescue. A sign at the bottom of the stairwell reads: "DETOUR: THIS STAIRCASE
   $[x = (t - 4)\cos t$, $y = (t - 4)\sin t$, $z = t^2$, $0 \leqslant t \leqslant 4]$ UNDER CONSTRUC-
   TION. USE ALTERNATE STAIRCASE — $(x = 0$, $y = -4\cos t$, $z = -4|\sin 2t|)$."

   (a) Sketch the path of the old route.

   (b) At $t = \pi$, another sign reads: "EXIT TO HOSTAGE ROOM. USE $x = 0$,
   $y = 4 - 4t$, $z = 16t$, $0 \leqslant t \leqslant 1$." Finally, when Prince Charming
   saves Cute Karen, another sign magically appears. It says: "HA! HA!
   THE PATH HAS COLLAPSED BEHIND YOU. YOU MAY GET TO THE GROUND BY REFLEC-
   TING YOUR ROUTE ACROSS THE PLANE $y = 0$. SIGNED, MEAN MAGICIAN."
   Sketch the entire rescue route.

ANSWERS TO PREREQUISITE QUIZ

1.   (a)

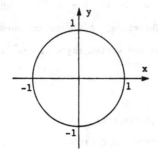

      (b)

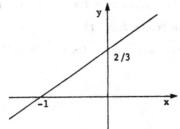

2.   (a) $(3x^2 + 1)(x^2 + 1) + (x^3 + x)(2x) = 5x^4 + 6x^2 + 1$

      (b) $-1/(x - 1)^2$

      (c) $3 \cos(3t - 1)$

ANSWERS TO SECTION QUIZ

1.

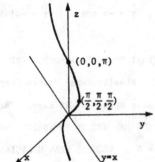

2. $\underline{\sigma}'(t) = (\sin t + t \cos t, \sin t + t \cos t, 1)$; $\underline{\sigma}''(t) = (2 \cos t - t \sin t,$
      $2 \cos t - t \sin t, 0)$

3.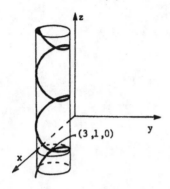

      (3,1,0)

4.   (a) $3 \ln t + 3 + (1/\sqrt{1 - t^2} - \sin^{-1}t)e^t - t \sin t/(1 - t) + \cos t/(1 - t)^2$

      (b) $[te^t/(1 - t) - (\sin^{-1}t)(\sin t) - \cos t/\sqrt{1 - t^2} + e^t/(1 - t)^2]\underline{i} +$
           $[(\ln t)(\sin t) - 3t/(1 - t) - 3t/(1 - t)^2 + \cos t/t]\underline{j} + [-3 \sin^{-1}t -$
           $(e^t)(\ln t) - e^t/t + 3t/\sqrt{1 - t^2}]\underline{k}$

      (c) $6t\underline{i} + 2t \exp(t^2)\underline{j} - 2t \sin(t^2)\underline{k}$

      (d) $(3 - 1/y)\underline{i} + (e^y - 1/\sqrt{1 - y^2})\underline{j} + (1/(1 - t)^2 - \sin t)\underline{k}$

5.   (a)

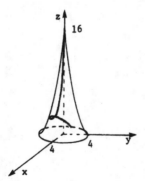

5.   (b)

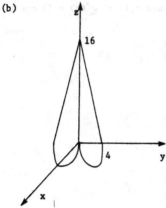

14.7 The Geometry and Physics of Space Curves

PREREQUISITES

1. Recall how to compute the length of a curve in the plane described by parametric equations (Section 10.4).

2. Recall Newton's second law (Section 8.1).

PREREQUISITE QUIZ

1. If a curve is described parametrically by $x = t^3 + 2t$ and $y = t^2 + 3$, what is the length of the curve for $0 \leq t \leq 2$? Express your answer as a definite integral.

2. State Newton's second law.

GOALS

1. Be able to compute the arc length of a vector function.

2. Be able to compute the curvature of a curve.

3. Be able to solve problems involving Newton's second law for curves.

STUDY HINTS

1. Arc length. In Section 10.4, you were given a formula for the arc length of a parametric curve in a plane. For the length of a curve in space, we simply add $(h'(t))^2$ under the radical sign, as you might have expected. Another formula for arc length is $L = \int_a^b \|\underline{\sigma}'(t)\| \, dt$. Thinking of $\|\underline{\sigma}'(t)\|$ as speed, the arc length equation becomes: distance $= \int_a^b (\text{speed}) dt$.

2. Arc length integration. A very useful formula is $\int \sqrt{x^2 + a^2} \, dx = (1/2) [x\sqrt{x^2 + a^2} + a^2 \ln(x + \sqrt{x^2 + a^2})] + C$. Keep this in mind for reference, but don't try to memorize it. (Many instructors will state it on an exam or give you a short table of integrals. Ask.) See how it is used in Example 2.

3. <u>Planetary motion</u>. The main point that you should get out of the dis-
    cussion about Newton's and Kepler's findings is that $T^2 = r_0^3 (2\pi)^2 / GM$.
    This is not to be memorized. It just shows how calculus can be applied
    to the real world.

4. <u>Curvature</u>. Curvature is a rate which tells you how fast the direction
    of motion is changing. A large curvature is associated with a sharp
    bend. A straight line has no curvature. And a small circle has large
    curvature. Since curvature is a rate, it is a derivative.

5. <u>Definitions</u>. Let $\underline{v}(t)$ be the velocity vector of a curve. If $\underline{v} \neq \underline{0}$
    for all t , the curve is called regular. $\underline{T} = \underline{v} / \|\underline{v}\|$ is the unit tan-
    gent vector, and if $\|\underline{v}\| = 1$ for all t , then the curve is said to
    be parametrized by arc length. Example 8 shows how regular curves can
    be converted to curves parametrized by arc length.

6. <u>Curvature formula</u>. This formula only applies to curves parametrized by
    arc length. It is $k = \|d\underline{T}/ds\|$, where s is the parameter we use
    rather than t when a curve is parametrized by arc length.

7. <u>More definitions</u>. The principal normal vector is $\underline{N} = (d\underline{T}/ds)/k = (d\underline{T}/ds)/\|d\underline{T}/ds\|$. It is orthogonal to $\underline{T}$ and points from the concave
    side of the curve. See Fig. 14.7.4.

8. <u>Supplement (Optional)</u>. The purpose of the supplement is to derive the
    sunshine formula (number 6 on p. 759) using vectors and rotations.
    Formula (6) and its derivation are not intended to be memorized.
    However, if you enjoy mathematics, you may like this derivation.
    Treat it as fun!

SOLUTIONS TO EVERY OTHER ODD EXERCISE

1. The arc length L is $\int_a^b [(dx/dt)^2 + (dy/dt)^2 + (dz/dt)^2]^{1/2} dt$. In

    this case, the length is $\int_0^{2\pi} [(-2 \sin t)^2 + (2 \cos t)^2 + 1]^{1/2} dt =$

    $\int_0^{2\pi} \sqrt{5} \, dt = 2\pi\sqrt{5}$.

5. The length is $\int_1^2 \sqrt{1 + 1 + (2t)^2} \, dt = \int_1^2 \sqrt{2 + 4t^2} \, dt = 2\int_1^2 \sqrt{1/2 + t^2} \, dt$.

    This has the form $\int \sqrt{x^2 + a^2}$, where $a = \sqrt{1/2}$. Thus, $L = 2(1/2) \times$

    $[t\sqrt{t^2 + 1/2} + (1/2) \ln (t + \sqrt{t^2 + 1/2})] \big|_1^2 = 2\sqrt{9/2} - \sqrt{3/2} + (1/2) \times$

    $[\ln (2 + \sqrt{9/2}) - \ln (1 + \sqrt{3/2})] = (6\sqrt{2} - \sqrt{6})/2 + (1/2) \ln [(2\sqrt{2} + 3)/$

    $(\sqrt{2} + \sqrt{3})] \approx 3.326$.

9. The period is given by $T = \sqrt{r_0^3 (2\pi)^2 / GM}$. Let R be the radius of the

    earth in meters. 500 miles $\approx 8.05 \times 10^5$ meters, so $T =$

    $\sqrt{(R + 8.05 \times 10^5)^3 (2\pi)^2 / GM}$. Using the constants from Example 5, we

    get $T = [(7.17 \times 10^6)^3 (2\pi)^2 / (6.67 \times 10^{-11})(5.98 \times 10^{24})]^{1/2} \approx$

    (6.50×10^3) seconds.

13. (a) $\underline{F} = m\underline{a} = m\underline{c}''(t) = m(x''\underline{i} + y''\underline{j} + z''\underline{k}) = (q/c)\underline{v} \times \underline{B} = [(q/c)(x'\underline{i} +$

        $y'\underline{j} + z'\underline{k})] \times b\underline{k} = (qb/c)y'\underline{i} - (qb/c)x'\underline{j}$. Equating the $\underline{i}, \underline{j}$,

        and $\underline{k}$ components, we get $x'' = (qb/cm)y'$; $y'' = -(qb/cm)x'$;

        $z'' = 0$.

    (b) Integrating $x'' = (qb/mc)y'$ yields $x' + C = (qb/mc)y$. When

        $t = 0$, $x' = 0$ and $y = 0$ by statement (2). Thus, $C = 0$.

        Substituting into the equation for y'' , we obtain $y'' =$

        $-(qb/mc)^2 y$, which is the spring equation with $\omega = qb/mc$. We

        have $y_0 = 0$ and $y_0' = a$, so $y = y_0 \cos \omega t + (y_0'/\omega) \sin \omega t =$

        $(amc/qb) \sin (qbt/mc)$.

            Substituting for y , we get $x' = a \sin(qbt/mc)$, and

        integration yields $x + C = -(amc/qb) \cos(qbt/mc)$. When $t = 0$,

        $x = 1$, so $C = -amc/qb - 1$; therefore, $x = -(amc/qb) \cos(qbt/mc) +$

13. (b) (continued)

   $amc/qb + 1$.

   Integration of $z'' = 0$ yields $z' = C$, but $z_0' = c$ implies $z' = c$. Integrating again gives $z = ct + C$. Since $z_0 = 0$, $z = ct$.

   (c) In the xy-plane, we have $(x - amc/qb - 1)^2 + y^2 =$ $(-amc/qb)^2\cos^2(qbt/mc) + (amc/qb)^2\sin^2(qbt/mc) = (amc/qb)^2$. This is a circle of radius amc/qb centered at $(amc/qb + 1, 0)$. The z-component describes the up and down motion, so the path is a right circular helix. The axis of the cylinder is parallel to the z-axis and passes through the point $(amc/qb + 1, 0, 0)$.

17. The ellipse $x^2 + 2y^2 = 1$, $z = 0$ can be parametrized by $\underline{\sigma}(t) =$ $(\cos t, \sin t/\sqrt{2}, 0)$. Then $\underline{v} = \underline{\sigma}'(t) = (-\sin t, (1/\sqrt{2})\cos t, 0)$ and $\underline{v}' = (-\cos t, -(1/\sqrt{2})\sin t, 0)$. Thus, the curvature is $k = \|\underline{v} \times \underline{v}\| /$ $\|\underline{v}\|^3 = \|(0, 0, (1/\sqrt{2})\sin^2 t + (1/\sqrt{2})\cos^2 t)\| / \|(-\sin t, (1/\sqrt{2})\cos t, 0)\|^3 =$ $1/\sqrt{2}(\sin^2 t + \cos^2 t/2)^{3/2} = 1/\sqrt{2}(2y^2 + x^2/2)^{3/2}$.

21. The particle is moving at constant speed, so let $\|\underline{v}\| = c$. Reparametrize by arc length using $s = ct$. Then $\underline{\sigma}(t) = \underline{\sigma}(s/c) = p(s)$, and so the unit tangent vector is $\underline{T} = p'(s) = (1/c)\underline{\sigma}'(s/c)$, where $\underline{\sigma}'(s/c)$ is the velocity. Then $d\underline{T}/ds = (1/c^2)\underline{\sigma}''(s/c)$, where $\underline{\sigma}''(s/c)$ is the acceleration. Now, by definition, $k = \|d\underline{T}/ds\|$ and $\underline{N} = (d\underline{T}/ds)/\|d\underline{T}/ds\|$. Therefore, the force is $\underline{F} = m\underline{a} = mc^2(d\underline{T}/ds)$, and so, $\|\underline{F}\| = mc^2 k$.

25. By definition, $\underline{N} = (d\underline{T}/ds)/\|d\underline{T}/ds\| = (d\underline{T}/ds)/k$. Rearrangement gives the first equation: $d\underline{T}/ds = k\underline{N}$. Next, using the chain rule, we get $d\underline{B}/ds = (d\underline{B}/dt)\cdot(dt/ds) = (-\tau\|\underline{v}\|\underline{N})\cdot(dt/ds)$. But $dt/ds = 1/(ds/dt) = 1/\|\underline{v}\|$, so $d\underline{B}/ds = (-\tau\|\underline{v}\|\underline{N})\cdot\|\underline{v}\| = -\tau\underline{N}$, which is the third equation.

25.   (continued).

dN/ds has the form $x\underline{T} + y\underline{N} + z\underline{B}$. Take the dot product with $\underline{T}$ to

get $(d\underline{N}/ds)\cdot\underline{T} = x(\underline{T}\cdot\underline{T}) + y(\underline{N}\cdot\underline{T}) + z(\underline{B}\cdot\underline{T})$. Since $\underline{B}$, $\underline{T}$, and $\underline{N}$

are mutually orthogonal unit vectors, we get $(d\underline{N}/ds)\cdot\underline{T} = x$. Similarly,

$(d\underline{N}/ds)\cdot\underline{B} = x(\underline{T}\cdot\underline{B}) + y(\underline{N}\cdot\underline{B}) + z(\underline{B}\cdot\underline{B}) = z$, and $(d\underline{N}/ds)\cdot\underline{N} = x(\underline{T}\cdot\underline{N}) +$

$y(\underline{N}\cdot\underline{N}) + z(\underline{B}\cdot\underline{N}) = y$. We also know that $0 = d(\underline{N}\cdot\underline{N})/ds = (d\underline{N}/ds)\cdot\underline{N} +$

$\underline{N}\cdot(d\underline{N}/ds) = 2\underline{N}\cdot(d\underline{N}/ds)$ or $(d\underline{N}/ds)\cdot\underline{N} = 0$, so $y = 0$. Also,

$0 = d(\underline{N}\cdot\underline{B})/ds = (d\underline{N}/ds)\cdot\underline{B} + \underline{N}\cdot(d\underline{B}/ds) = (d\underline{N}/ds)\cdot\underline{B} + \underline{N}\cdot(-\tau\underline{N}) = (d\underline{N}/ds)\cdot\underline{B} -$

$\tau\|\underline{N}\|^2$. Thus, $(d\underline{N}/ds)\cdot\underline{B} = \tau$, so $z = \tau$. And $0 = d(\underline{N}\cdot\underline{T})/ds =$

$(d\underline{N}/ds)\cdot\underline{T} + \underline{N}\cdot(d\underline{T}/ds) = (d\underline{N}/ds)\cdot\underline{T} + \underline{N}\cdot(k\underline{N}) = (d\underline{N}/ds) + k\|\underline{N}\|^2$, so

$x = -k$. Therefore, $d\underline{N}/ds = -k\underline{T} + \tau\underline{B}$.

SECTION QUIZ

1.   Find the arc length of the following vector functions:

   (a) $(3t^4 + 3,8\sqrt{2}t^3, 24t^2 - 13)$, $-1 \leqslant t \leqslant 1$

   (b) $(-2,5,1)$, $2 \leqslant t \leqslant 4$

   (c) $6t^2\underline{i} + 9\underline{j} - 5t\underline{k}$ for $0 \leqslant t \leqslant 3$

2.   (a) Suppose the curvature of a vector function is zero, what can you say
         about the curve?

     (b) Suppose the curvature of a vector function was found to be negative,
         what can you say about the curve?

3.   A ghostly planet has been discovered revolving around the sun at a distance
     which is twice the Earth-sun distance. An invisibility shield had kept the
     planet a secret until radar discovered a mass equal to the Earth's at the
     specified location. How many Earth years does the ghost planet need to
     orbit the sun?

4.   Reckless Roger, the famous car racer, suffers from curviphobia (the fear
of curves). He refuses to race on any track which has curvature greater
than 2 . Suppose a track is laid down along the curve parametrized by
$(x,y,z) = (\cos t, 2 \sin t, 0)$.

(a) Sketch the race track.

(b) Using the sketch in (a) and your knowledge of curvature, find the
points where the curvature is smallest.

(c) Find the maximum curvature to determine if Reckless Roger will suffer
curviphobia on this race track.

ANSWERS TO PREREQUISITE QUIZ

1. $\int_0^2 (9t^4 + 16t^2 + 4)^{1/2} dt$

2. $\underline{F} = m(d^2\underline{x}/dt^2)$

ANSWERS TO SECTION QUIZ

1.   (a) 54

(b) 0

(c) $3\sqrt{1321}/2 + (25/12) \ln |(36 + \sqrt{1321})/5|$

2.   (a) It is a straight line.

(b) Nothing; an error was made. Curvature is never negative.

3. $2\sqrt{2}$ years

4.   (a)

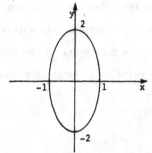

4.   (b) (±1,0)

(c) Maximum curvature at $(0,\pm 2) = 2$, so the track is barely
acceptable.

14.S <u>Supplement to Chapter 14: Rotations and the Sunshine Formula</u>
SOLUTIONS TO EVERY OTHER ODD EXERCISE

1. (a) The angle between $\underline{1}$ and $\underline{r}_0$ is given by $\cos \lambda = \underline{1} \cdot \underline{r}_0 = -1/2$.
     $\lambda = 2\pi/3$ and $\sin \lambda = \sqrt{3}/2$. From formula (1), we get $\underline{m}_0 =$
     $\underline{r}_0/\sin \lambda - \underline{1} \cos \lambda/\sin \lambda = (2/\sqrt{3})(1/\sqrt{2})(\underline{j} + \underline{k}) - (1/\sqrt{3})(1/\sqrt{2})(\underline{i} - \underline{j}) =$
     $(1/\sqrt{6})(\underline{i} + \underline{j} + 2\underline{k})$, and $\underline{n}_0 = \underline{1} \times \underline{m}_0 = (1/2\sqrt{3})(\underline{i} + \underline{j} - \underline{k})$.

   (b) From formula (2), $\underline{r} = \underline{\sigma}(t) = \cos \lambda \underline{1} + \sin \lambda \cos (2\pi t/T)\underline{m}_0 +$
     $\sin \lambda \sin(2\pi t/T)\underline{n}_0 = (-1/2)(\underline{j} + \underline{k})/\sqrt{2} + (\sqrt{3}/2)\cos(\pi t/12)(1/\sqrt{6}) \times$
     $(\underline{i} + \underline{j} + 2\underline{k}) + (\sqrt{3}/2) \sin(\pi t/12)(1/2\sqrt{3})(\underline{i} + \underline{j} - \underline{k}) = [\cos(\pi t/12)/$
     $2\sqrt{2} + \sin(\pi t/12)/4]\underline{i} + [-1/2\sqrt{2} + \cos(\pi t/12)/2\sqrt{2} + \sin(\pi t/12)/4]\underline{j} +$
     $[-1/2\sqrt{2} + \cos(\pi t/12)/\sqrt{2} - \sin(\pi t/12)/4]\underline{k}$.

   (c) When $t = 12$ and $T = 24$, part (b) tells us that $\underline{\sigma}(12) =$
     $(-1/2\sqrt{2})\underline{i} - (1/\sqrt{2})\underline{j} - (3/2\sqrt{2})\underline{k}$. $\underline{\sigma}'(t) = [-(\pi/12)\sin(\pi t/12)/2\sqrt{2} +$
     $(\pi/12)\cos(\pi t/12)/4]\underline{i} + [-(\pi/12)\sin(\pi t/12)/2\sqrt{2} + (\pi/12)\cos(\pi t/12)/$
     $4]\underline{j} + [-(\pi/12)\sin(\pi t/12)/\sqrt{2} - (\pi/12)\cos(\pi t/12)/4]\underline{k}$, so $\underline{\sigma}'(12) =$
     $(-\pi/48)\underline{i} - (\pi/48)\underline{j} + (\pi/48)\underline{k}$. Therefore, the tangent line is
     $(-1/2\sqrt{2})(\underline{i} + 2\underline{j} + 3\underline{k}) + (-\pi/48)(\underline{i} + \underline{j} + \underline{k})(t - 12)$.

5. Sunset occurs when $\sin A = 0$. Let $B = 2\pi t/T_y$ and $C = 2\pi t/T_d$.
   Thus, we have $-\cos B(\sin \ell \sin \alpha + \cos \ell \cos \alpha \cos C) =$
   $\sin B \cos \ell \sin C$. Dividing by $-\cos B \cos \ell$ gives $\tan \ell \sin \alpha +$
   $\cos \alpha \cos C = -\tan B \sin C$, i.e., $\tan \ell \sin \alpha = -\tan B \sin C -$
   $\cos \alpha \cos C = -\cos C(\tan B \tan C - \cos \alpha)$. Thus, our "exact" formula
   is $-\tan \ell \sin \alpha = \cos(2\pi t/T_d) [\tan(2\pi t/T_y)\tan(2\pi t/T_d) - \cos \alpha]$.

9. When $\alpha = 32°$, E for Paris is 0.1289 and E at the equator is
   0.8480 . Thus, the equator receives more than 6 times as much solar
   energy on a summer day than Paris receives on January 15, provided that
   the earth's tilt has changed.

14.R Review Exercises for Chapter 14

SOLUTIONS TO EVERY OTHER ODD EXERCISE

1.

$4x^2 + 9y^2 = 36$ is the same as $x^2/9 + y^2/4 = 1$. This is an ellipse with intercepts at $x = \pm 3$ and $y = \pm 2$.

5.

$100x^2 + 100y^2 = 1$ is equivalent to $x^2 + y^2 = 1/100$, which is a circle centered at the origin with radius $1/10$.

9.

By completing the squares, we have $(9x^2 - 18x + 9) + (y^2 - 4y + 4) = 9 = 9(x - 1)^2 + (y - 2)^2$ or $(x - 1)^2 + (y - 2)^2/9 = 1$. This has the form $X^2 + Y^2/9 = 1$, which is an ellipse with intercepts at $X = \pm 1$ and $Y = \pm 3$. Shifting this ellipse's center to $(1,2)$ gives the desired graph.

13.

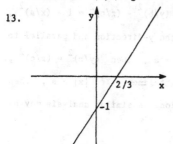

$c = 2 = 3x - 2y$ implies $2y = 3x - 2$ or $y = 3x/2 - 1$. This is a line with slope $3/2$ and y-intercept at -1.

17.

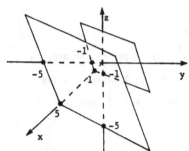

$c = x - y - z$ is a plane for any value of c. These planes are parallel, whose position depends on the values of c.

21.

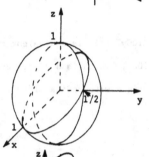

This is an ellipsoid centered at the origin with intercepts $(\pm 1,0,0)$, $(0,\pm 1/2,0)$ and $(0,0,\pm 1)$.

25.

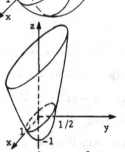

$x^2 + 4y^2 = 1 + z$ is an ellipse parallel to the xy-plane when z is held constant if $z > -1$. This is an elliptic paraboloid with intercepts $(\pm 1,0,0)$, $(0,\pm 1/2,0)$, and $(0,0,-1)$.

29. (a) $(x/a)^2 + (y/b)^2 = 1 + (z/c)^2$ is an ellipse in any plane parallel to the xy-plane when z is held constant.

(b) When x is held constant, we have $(y/b)^2 - (z/c)^2 = 1 - (x/a)^2$, which is a hyperbola opening along the y-direction and parallel to the yz-plane if $|x| < a$. If $|x| = a$, then $(y/b)^2 = (z/c)^2$, or $y = \pm bz/c$, which is two straight lines. If $|x| > a$, then the hyperbola opens in the z-direction. A similar analysis may be done for constant y.

29. (c)

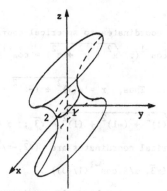

The easiest way to sketch this is to draw ellipses parallel to the xy-plane and enlarge them as $|z|$ increases.

In the yz-plane, $x = 0$ gives the equation $y^2 - z^2 = 1$ which is a hyperbola with y-intercepts, ± 1 . For $x = 1$, the equation is $y^2 - z^2 = 3/4$, so the y-intercepts are $\pm\sqrt{3}/2$ for this hyperbola. When $x = 2$, $y^2 = z^2$ or $y = |z|$, which is two straight lines. And $x = 3$ yields the equation $y^2 - z^2 = -5/4$. This is a hyperbola with z-intercepts, $\pm\sqrt{5}/2$.

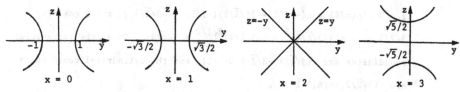

33.

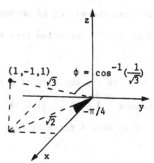

To convert cylindrical coordinates to rectangular coordinates, use $x = r\cos\theta$, $y = r\sin\theta$, $z = z$. To convert from rectangular coordinates to cylindrical coordinates, use $r = \sqrt{x^2 + y^2}$,

$$\theta = \begin{cases} \tan^{-1}(y/x) & \text{if } x \geq 0 \\ \tan^{-1}(y/x) + \pi & \text{if } x < 0 \end{cases},$$

and $z = z$.

33. (continued)

The conversion from rectangular coordinates to spherical coordinates

uses $\rho = \sqrt{x^2 + y^2 + z^2}$, $\phi = \cos^{-1}(z/\sqrt{x^2 + y^2 + z^2}) = \cos^{-1}(z/\rho)$,

$$\theta = \begin{cases} \tan^{-1}(y/x) & \text{if } x \geqslant 0 \\ \tan^{-1}(y/x) + \pi & \text{if } x < 0 \end{cases} . \quad \text{Thus,} \quad r = \sqrt{(1)^2 + (-1)^2} = \sqrt{2} ;$$

$\theta = \tan^{-1}(-1/1) = -\pi/4$; $\rho = \sqrt{(1)^2 + (-1)^2 + (1)^2} = \sqrt{3}$; $\phi =$

$\cos^{-1}(1/\sqrt{3})$. Thus, the cylindrical coordinates are $(\sqrt{2},-\pi/4,1)$ and

the spherical coordinates are $(\sqrt{3},-\pi/4,\cos^{-1}(1/\sqrt{3}))$.

37.

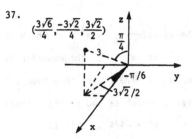

To convert spherical coordinates to rectangular coordinates, use $x =$
$\rho \sin\phi \cos\theta$, $y = \rho \sin\phi \sin\theta$, $z =$
$\rho \cos\phi$. After converting to rectangular
coordinates, r is obtained from
$\sqrt{x^2 + y^2}$. θ is the same as the spherical

coordinate. Thus, $x = 3\sin(\pi/4)\cos(-\pi/6) = 3(\sqrt{2}/2)(\sqrt{3}/2) = 3\sqrt{6}/4$;

$y = 3\sin(\pi/4)\sin(-\pi/6) = 3(\sqrt{2}/2)(-1/2) = -3\sqrt{2}/4$; $z = 3\cos(\pi/4) =$

$3\sqrt{2}/2$; $r = [(3\sqrt{6}/4)^2 + (-3\sqrt{2}/4)^2]^{1/2} = 3\sqrt{2}/2$. Thus, the rectangular

coordinates are $(3\sqrt{6}/4,-3\sqrt{2}/4,3\sqrt{2}/2)$ and the cylindrical coordinates

are $(3\sqrt{2}/2,-\pi/6,3\sqrt{2}/2)$.

41. Replacing (ρ,θ,ϕ) by $(\rho,\theta + \pi,\phi + \pi/2)$ has the effect of moving each

point $180°$ around the z-axis, followed by a $90°$ rotation from the

positive z-axis.

45.

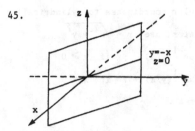

$x + y - z = 0$ is the equation of a plane.

Plot the points $(0,0,0)$, $(1,0,1)$ and

$(0,1,1)$. Then draw a plane through these

points.

49.

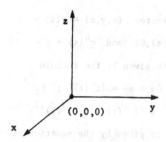

$-(x^2 + y^2/4)$ is nonpositive and z^2 is nonnegative; therefore, the only solution is $(0,0,0)$.

53. The equation of the tangent line is $(f(t_0),g(t_0),h(t_0)) + (t - t_0) \times$ $(f'(t_0),g'(t_0),h'(t_0))$. In this case, $(f'(t),g'(t),h'(t)) = (3t^2,$ $-e^{-t},-(\pi/2) \sin(\pi t/2))$, so the tangent line is $(2,1/e,0) + (t - 1) \times$ $(3,-1/e,-\pi/2)$.

57. The velocity vector of $f(t)\underline{i} + g(t)\underline{j} + h(t)\underline{k}$ is $f'(t)\underline{i} + g'(t)\underline{j} +$ $h'(t)\underline{k}$ and the acceleration vector is $f''(t)\underline{i} + g''(t)\underline{j} + h''(t)\underline{k}$. By the sum rule, $\underline{\sigma}'(t) = \underline{v}(t) = \underline{\sigma}_1'(t) + \underline{\sigma}_2'(t) = [e^t + 2t/(1 + t^2)^2]\underline{i} +$ $(\cos t + 1)\underline{j} - \sin t \, \underline{k}$; $\underline{\sigma}''(t) = \underline{a}(t) = \underline{\sigma}_1''(t) + \underline{\sigma}_2''(t) = [e^t +$ $(2 - 6t^2)/(1 + t^2)^3]\underline{i} - \sin t \, \underline{j} - \cos t \, \underline{k}$.

61. The arc length is $L = \int_1^2 [(dx/dt)^2 + (dy/dt)^2 + (dz/dt)^2]^{1/2}dt =$ $\int_1^2 (1 + 1/t^2 + 2/t)^{1/2}dt = \int_1^2 [(t^2 + 1 + 2t)/t^2]^{1/2}dt = \int_1^2 [(t + 1)/t]dt =$ $(t + \ln t)\big|_1^2 = 1 + \ln 2$.

65. (a) Let $\underline{r}(t) = a(t)\underline{i} + b(t)\underline{j} + c(t)\underline{k}$. $\underline{F} = m\underline{a} = m\underline{r}''(t) = -k\underline{r}(t)$, so rearrangement gives $\underline{r}''(t) = -(k/m)\underline{r}(t)$. Thus, the differential equations are $a''(t) = -(k/m)a(t)$, $b''(t) = -(k/m)b(t)$, and $c''(t) = -(k/m)c(t)$.

(b) From the solution of the spring equation, we have $a = a_0\cos \omega t +$ $(a_0'/\omega) \sin \omega t$, where $\omega = k/m$. $\underline{r}(0) = \underline{0}$ implies $a_0 = b_0 =$ $c_0 = 0$ and $\underline{r}'(0) = 2\underline{j} + \underline{k}$ implies $a_0' = 0$, $b_0' = 2$, and $c_0' = 1$. Thus, $a(t) = 0$. Similarly, $b(t) = (2m/k) \sin (k/m)t$, and $c(t) = (m/k) \sin (k/m)t$.

69. Let the curve be parametrized by the vector $(x,y,z) = \underline{g}(t) =$
$(t,f(t),0)$. Then $\underline{g}'(t) = \underline{v} = (1,f'(t),0)$ and $\underline{g}''(t) = \underline{a} =$
$(0,f''(t),0)$. Recall that curvature is given by the formula
$\|\underline{v} \times \underline{a}\|/\|\underline{v}\|^3$. $\underline{v} \times \underline{a} = f''(t)\underline{k}$, so $\|\underline{v} \times \underline{a}\| = |f''(t)|$; $\|\underline{v}\|^3 =$
$(1 + |f'(t)|^2)^{3/2}$, so the curvature is $|f''(t)(1 + |f'(t)|^2)^{-3/2}|$.

73. (a) The intersection of the surfaces is given by the equation $x^2 +$
$1 + z^2 = 3$ or $x^2 + z^2 = 2$, which is a circle of radius $\sqrt{2}$
centered at $(0,1,0)$ in the plane $y = 1$. One parametric form
of the curve is $x = \sqrt{2} \cos t$, $y = 1$, $z = \sqrt{2} \sin t$. (There
are other solutions to this exercise.)

   (b) Using the answer we got in part (a), the velocity vector is
$(-\sqrt{2} \sin t, 0, \sqrt{2} \cos t)$. The curve is at $(1,1,1)$ when $t_0 =$
$\pi/4$, so the tangent line is $(1,1,1) + t(-1,0,1)$.

   (c) The curve is self-intersecting, so restrict the interval to
$0 \leqslant t \leqslant 2\pi$. The arc length is $\int_0^{2\pi} \sqrt{(-\sqrt{2} \sin t)^2 + (0)^2 + (\sqrt{2} \cos t)^2} dt$
$\int_0^{2\pi} \sqrt{2} \, dt = \sqrt{2} t \big|_0^2 = 2\sqrt{2}\pi$.

TEST FOR CHAPTER 14

1. True or false.

   (a) The derivative of $\underline{u} \times \underline{v} = \underline{u}' \times \underline{v} + \underline{v}' \times \underline{u}$.

   (b) The graph of the unit sphere is the graph of a function of x and y .

   (c) The derivative of $3\underline{i} + 2\underline{j} - \underline{k}$ is a scalar, namely 0 .

   (d) If the spherical coordinate ϕ is $5\pi/8$, then the point (ρ,θ,ϕ)
always lies below the xy-plane, provided $\rho > 0$.

   (e) Any equation of the form $Ax^2 + Bxy + Cy^2 + Dx + Ey + F = 0$, where
A , B , C , D , E , and F are constants, describes a conic section

2. A sphere of radius 2 is centered at $(-2,-1,0)$.

 (a) Write the equation of the sphere in spherical coordinates.

 (b) What are the cylindrical coordinates of the sphere's center?

3. (a) Let $\underline{w}(t) = (t^5 + t)\underline{i} - (\sin t)\underline{j}$. What is the derivative of $\underline{w}(x^3 + 5x^2 - x + 1)$ with respect to x ?

 (b) If $\underline{u}$, $\underline{v}$, and $\underline{r}$ are vector functions, find a formula for $(d/dt)[\underline{u}(t) \cdot \underline{r}(t)]\underline{v}(t)$.

4. (a) Sketch the graph of $2y = x$ in xyz-space.

 (b) Using the definition of a cylinder, explain why the surface in part (a) is or is not a cylinder.

5. 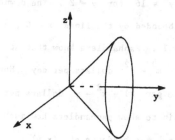 The surface shown at the left is a cone (not of the elliptical type).

 (a) Write the general equation of the surface.

 (b) Describe the level curves for y = constant.

6. (a) Sketch the parametric curve $(e^t, e^{2t}, 2e^t)$, $0 \leq t \leq 2$.

 (b) Compute the length of the curve sketched in part (a).

 (c) What is the velocity vector for the parametric curve?

7. Describe the graphs of the following:

 (a) θ = constant in cylindrical coordinates.

 (b) $\phi = \pi/4$, $\rho \leq 0$ in spherical coordinates.

 (c) $\theta = 30°$, $\phi = \tan^{-1} 2$ in spherical coordinates.

8. (a) Sketch the surface described by the equation $4x^2 + 4y^2 + z^2 = 1$.

    (b) If the origin is shifted to $(-1,0,2)$, what is the equation of the new surface?

    (c) If an insect crawls along the original surface from $(0,0,1)$ to $(1/4, 1/4, \sqrt{11}/4)$ along the shortest path where $x = y$, how far does it travel? (Leave your answer in terms of an integral.)

9. (a) Compute the curvature of the circle $x^2 + y^2 = 9$.

    (b) What is the principal normal vector to the circle at $(0,1)$ if the circle is traced out in a counterclockwise direction?

10.
A certain city's boundary, as depicted, is given by the circle $x^2 + y^2 = 16$ for $y \geqslant 2$. The downtown area (shaded) is bounded by the lines $x = 0$, $x = 2$, $y = -1$, and $y = 1$. Panhandlers know that at (x,y), they can get $16 - x^2 - y^2$ dollars per day. However, in the downtown area, they expect to get $21 - x^2 - y^2$ dollars per day. Sketch a graph over the city's domain to show panhandlers how much they can expect to receive if they ask for handouts at (x,y).

ANSWERS TO CHAPTER TEST

1. (a) False; it is $\underline{u}' \times \underline{v} + \underline{u} \times \underline{v}'$.

    (b) False; there are two values for z for most (x,y) where the graph exists.

    (c) False; the derivative is the vector, $\underline{0}$.

    (d) True

    (e) False; only if the graph exists.

2. (a) $\rho^2 + \rho \sin \phi (4 \cos \theta + 2 \sin \theta) + 1 = 0$

    (b) $(\sqrt{5}, -5\pi/6, 0)$

3. (a) $(3x^2 + 10x - 1)[(5(x^3 + 5x^2 - x + 1)^4 + 1)\underline{i} - (\cos(x^3 + 5x^2 - x + 1))\underline{j}]$

(b) $[\underline{u}'(t)\cdot\underline{r}(t) + \underline{u}(t)\cdot\underline{r}'(t)]\underline{v}(t) + [\underline{u}(t)\cdot\underline{r}(t)]\underline{v}'(t)$

4. (a)

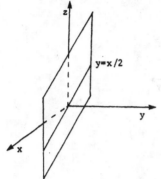

(b) The surface is a cylinder because z does not appear in the equation.

5. (a) $x^2 + z^2 = y$; $y \geqslant 0$

(b) The level curves are circles of radius $\sqrt{y}$, centered on the y-axis
if $y \geqslant 0$.

6. (a)

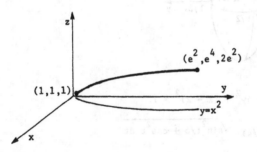

(b) $[2e^2\sqrt{4e^4 + 5} - 6 + 5 \ln (2e^2 + \sqrt{4e^4 + 5}) - 5 \ln 5]/4$

(c) $(e^t, 2e^{2t}, 2e^t)$

7.   (a) xz-plane rotated θ radians.

(b)   A cone lying below the xy-plane.

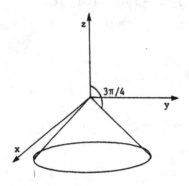

(c) A straight line.

8.   (a)

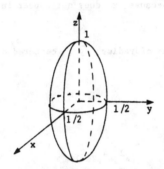

(b) $(x + 1)^2 + 4y^2 + (z - 2)^2 = 1$

(c) $\int_{\sin^{-1}(\sqrt{11}/4)}^{\pi/2} \sqrt{\sin^2 t/5 + \cos^2 t} \; dt$

9.   (a) 1/3

(b) $-\underline{i}$

10.

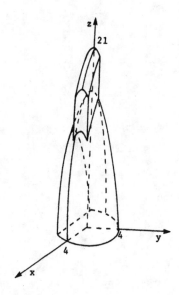

CHAPTER 15

PARTIAL DIFFERENTIATION

15.1 Introduction to Partial Derivatives

PREREQUISITES

1. Recall the basic rules of one-variable differentiation (Chapters 1 and
    2, Sections 5.2, 5.4, and 6.3).

2. Recall the definition of the derivative for one variable (Section 1.3).

3. Recall the definition of continuity with one variable (Section 3.1).

PREREQUISITE QUIZ

1. Differentiate the following:

    (a) $(\sqrt{x} + 1)^2 (x^3 + x - 3)$

    (b) $x/(2 + \cos x)^2$

    (c) $\exp(\tan x)$

2. If $f(x) = x^3 + 2$, what is $[df^{-1}(x)/dx]\big|_{10}$?

3. Use limits to prove that $(d/dx)x^2 = 2x$.

4. State two conditions that must be met for a one-variable function to be
    called continuous.

GOALS

1. Be able to state the definition of a partial derivative.

2. Be able to compute partial derivatives.

3.   Be able to explain in "common terms" what partial derivatives tell you.

4.   Be able to explain the relationship between mixed partials.

STUDY HINTS

1.   <u>Partial derivatives</u>. The three main notations used to represent partials
     are f_x , $\partial z/\partial x$, and $\partial f/\partial x$. Differentiation is performed with re-
     spect to the variable in the subscript or the "denominator". In this
     case, all independent variables in the function f are held constant
     except for x and differentiation is performed as usual.

2.   <u>Definition</u>. The definition of partial derivatives is almost the same as
     the derivative in one-variable calculus. Remember that only x is
     changing in $\partial f/\partial x$, so $\partial f/\partial x$ is $\lim\limits_{\Delta x \to 0}\{[f(x + \Delta x,y,z) - f(x,y,z)]/\Delta x\}$.

3.   <u>Equality of mixed partials</u>. This concept allows you to calculate second
     or higher partials in any order as long as you differentiate with respect
     to each variable the specified number of times. You need continuous par-
     tial derivatives to apply this rule.

4.   <u>Evaluating partials at a given point</u>. Always remember to differentiate
     completely before substituting given values. With mixed partials, you
     may substitute for a variable only after you have completed differenti-
     ating in that variable.

5.   <u>Limits</u>. Example 9 shows that we need to get the same value in all direc-
     tions (not just along directions parallel to the x- and y-axes) if the
     limit exists. To show that a limit does not exist, we can often compute
     values in directions parallel to the coordinate axes and see if they are
     unequal.
     With the ε-δ definition, we're interested in what happens near a point.
     In two dimensions, nearness is defined by surrounding a point by a disk,
     which represents δ . A ball is used in three dimensions.

6. <u>Continuity</u>. As with one variable, the function is continuous wherever
    the limit equals the function value.

SOLUTIONS TO EVERY OTHER ODD EXERCISE

1. $f_x = y$, so $f_x(1,1) = 1$; $f_y = x$, so $f_y(1,1) = 1$.

5. By the product rule, $f_x = ye^{xy}\sin(x + y) + e^{xy}\cos(x + y)$, so
    $f_x(0,0) = 0 + \cos 0 = 1$; $f_y = xe^{xy}\sin(x + y) + e^{xy}\cos(x + y)$, so
    $f_y(0,0) = 1$. Note that, due to the symmetry of the function, f_y
    could have been found by reversing the roles of x and y .

9. $f_x = yz$, so $f_x(1,1,1) = 1$; $f_y = xz$, so $f_y(1,1,1) = 1$; $f_z =$
    xy , so $f_z(1,1,1) = 1$.

13. $\partial z/\partial x = 6x$ and $\partial z/\partial y = 4y$.

17. $\partial u/\partial x = [-yz \exp(-xyz)](xy + xz + yz) + [\exp(-xyz)](y + z) = \exp(-xyz) \times$
    $[-yz(xy + xz + yz) + (y + z)]$; $\partial u/\partial y = [-xz \exp(-xyz)](xy + xz + yz) +$
    $[\exp(-xyz)](x + z) = \exp(-xyz)[-xz(xy + xz + yz) + (x + z)]$; $\partial u/\partial z =$
    $[-xy \exp(-xyz)](xy + xz + yz) + [\exp(-xyz)](x + y) = \exp(-xyz) \times$
    $[-xy(xy + xz + yz) + (x + y)]$.

21. The partial is $[xe^y(ye^x + 1) - (xe^y - 1)e^x]/(ye^x + 1)^2 = (xye^xe^y -$
    $xe^xe^y + xe^y + e^x)/(ye^x + 1)^2$.

25. $f_x = 6x + (2/y^2)\cos(x/y^2) + y^3(-e^x)$, so $f_x(2,3) = 12 + (2/9)\cos(2/9) - 27e^2$

29. (a) $\partial z/\partial y = (\sin x)e^{-xy}(-x) = -x(\sin x)e^{-xy}$.

    (b) Evaluating $\partial z/\partial y$ at the given points, we have $z_y(0,0) = 0$;
         $z_y(0,\pi/2) = 0$; $z_y(\pi/2,0) = -\pi/2$; $z_y(\pi/2,\pi/2) = -(\pi/2)\exp(-\pi^2/4)$.

33. $g_u(t,u,v) = 1/(t + u + v) - (tv)\sec^2(tuv)$, so $g_u(1,2,3) = 1/6 -$
    $3 \sec^2(6)$.

37. $(\partial/\partial s)\exp(stu^2) = (tu^2)\exp(stu^2)$.

41. Using the idea of Example 2(c), we get $f_z = \lim_{\Delta z \to 0} \{ [f(x,y,z+\Delta z) -$
    $f(x,y,z)]/\Delta z \}$.

45. (a) $R = 1/(1/R_1 + 1/R_2 + 1/R_3)$, so $\partial R/\partial R_1 = -1/(1/R_1 + 1/R_2 + 1/R_3)^2(-1/R_1^2) = 1/(1 + R_1/R_3 + R_1/R_3)^2$.

    (b) $(\partial R/\partial R_1)|_{(100,200,300)} = 1/(1 + 1/2 + 1/3)^2 = (11/6)^{-2} = 36/121$,

    so R is changing $36/121$ times as fast.

49. From Exercise 15, we get $z = 2x/3y + 7x/3$, so $\partial z/\partial x = 2/3y + 7/3$

and $\partial z/\partial y = -2x/3y^2$. Thus, $\partial^2 z/\partial x^2 = (\partial/\partial x)(\partial z/\partial x) = 0$;

$\partial^2 z/\partial x\partial y = (\partial/\partial x)(\partial z/\partial y) = (\partial/\partial y)(\partial z/\partial x) = \partial^2 z/\partial y\partial x = -2/3y^2$;

$\partial^2 z/\partial y^2 = (\partial/\partial y)(\partial z/\partial y) = 4x/3y^3$.

53. $\partial u/\partial x = [(x^2 + y^2)^2(2y) - 2xy(4x)(x^2 + y^2)]/(x^2 + y^2)^4 = (-6x^2y + 2y^3)/$

$(x^2 + y^2)^3$; $\partial u/\partial y = [(x^2 + y^2)^2(2x) - 2xy(4y)(x^2 + y^2)]/(x^2 + y^2)^4 =$

$(-6xy^2 + 2x^3)/(x^2 + y^2)^3$; so $\partial^2 u/\partial x^2 = [(x^2 + y^2)^3(-12xy) -$

$(-6x^2y + 2y^3)(6x)(x^2 + y^2)^2]/(x^2 + y^2)^6 = 24xy(x^2 - y^2)/(x^2 + y^2)^4$;

$\partial^2 u/\partial y\partial x = (\partial/\partial y)(\partial u/\partial x) = [(x^2 + y^2)^3(-6x^2 + 6y^2) - (-6x^2y + 2y^3)(6y) \times$

$(x^2 + y^2)^2]/(x^2 + y^2)^6 = -6(x^4 - 6x^2y^2 + y^4)/(x^2 + y^2)^4$; $\partial^2 u/\partial x\partial y =$

$(\partial/\partial x)(\partial u/\partial y) = [(x^2 + y^2)^3(-6y^2 + 6x^2) - (-6xy^2 + 2x^3)(6x)(x^2 + y^2)^2]/$

$(x^2 + y^2)^6 = -6(x^4 - 6x^2y^2 + y^4)/(x^2 + y^2)^4 = \partial^2 u/\partial y\partial x$; and $\partial^2 u/\partial y^2 =$

$[(x^2 + y^2)^3(-12xy) - (-6xy^2 + 2x^3)(6y)(x^2 + y^2)^2]/(x^2 + y^2)^6 =$

$-24xy(x^2 - y^2)/(x^2 + y^2)^4$.

57. For any given $\varepsilon > 0$, take $\delta = \varepsilon$. Then, for any (x,y) such that

$0 < d((x,y),(x_0,y_0)) < \delta$, i.e., $\sqrt{(x - x_0)^2 + (y - y_0)^2} < \delta$, we

have $|y - y_0| < \delta$. Thus, $|y - y_0| < \varepsilon$, i.e., $|f(x,y) - \ell| < \varepsilon$,

also. So $\lim_{(x,y)\to(x_0,y_0)} y = y_0$.

61. Since the denominator doesn't vanish, the function is a "reasonable"

one in which the limit may be evaluated by substituting $(x,y) = (2,3)$.

Thus, the limit is $52/\sqrt{13} = 4\sqrt{13}$.

65. By direction substitution, $\lim_{(x,y)\to(0,0)} \sin(x,y) = \sin(0) = 0$.

69. If $u = x^3 - 3xy^2$, then $\partial u/\partial x = 3x^2 - 3y^2$ and $\partial^2 u/\partial x^2 = 6x$. Also, $\partial u/\partial y = -6xy$ and $\partial^2 u/\partial y^2 = -6x$. Therefore, $\partial^2 u/\partial x^2 + \partial^2 u/\partial y^2 = 6x - 6x = 0$. This means that $u(x,y)$ is a harmonic function.

73. (a) $(\partial z/\partial x)\big|_{(5,3)} = (60y - 2x)\big|_{(5,3)} = 180 - 10 = 170$ units.

   (b) $(\partial z/\partial y)\big|_{(5,3)} = (60x - 8y)\big|_{(5,3)} = (300 - 24) = 276$ units. This is the marginal productivity of capital, per million dollars invested, at a labor force of 5000 people and investment level of 3 million dollars.

77. (a) From Exercise 76(a), $f_x = y(x^4 + 4x^2y^2 - y^4)/(x^2 + y^2)^2$. Substitute $x = 0$ into f_x giving, for $y \neq 0$, $f_x(0,y) = y(-y^4)/y^4 = -y$.

   (b) From Exercise 76(a), $f_y = x(x^4 - 4x^2y^2 - y^4)/(x^2 + y^2)^2$. Substitute $y = 0$ into f_y giving, for $x \neq 0$, $f_y(x,0) = x(x^4)/x^4 = x$.

   (c) Using the definition of the derivative, $f_{yx}(0,0) = \lim_{\Delta x \to 0} \{[f_y(\Delta x,0) - f_y(0,0)]/\Delta x\} = \lim_{\Delta x \to 0} [(\Delta x - 0)/\Delta x] = \lim_{\Delta x \to 0} 1 = 1$. Similarly, $f_{xy}(0,0) = \lim_{\Delta y \to 0} \{[f_x(0,\Delta y) - f_x(0,0)]/\Delta y\} = \lim_{\Delta y \to 0} [(-\Delta y - 0)/\Delta y] = \lim_{\Delta y \to 0} (-1) = -1$.

   (d) By the equality of mixed partials, in order to have $f_{xy} = f_{yx}$, both f_{xx} and f_{yy} must be continuous. Without actually differentiating f_x or f_y (from Exercise 76(a)), we can see that each will have $(x^2 + y^2)^3$ in the denominator. Therefore, neither is defined, nor continuous, at $(0,0)$, so the equality of mixed partials does not apply.

SECTION QUIZ

1. Fill in the blanks: Suppose that f is a function of x, y, and z. f_x is the notation used to indicate differentiation with respect to _____ while the variable(s) _____ is (are) held constant.

2. If f_y is a partial derivative, write f_{yy} in Leibniz notation, and explain how to find f_{yy} from f_y.

3. (a) Write $\partial g/\partial z$ for a function $g(x,y,z)$ as a limit.

   (b) Using the limit method, find $g_z(0,1,1)$ for $g(x,y,z) = x^2 yz + xz^2 - xy$.

4. Find $\partial^3 g/\partial x^2 \partial y$ for $g(x,y) = \cos(e^x y)$.

5. (a) For the function in Question 4, does $\partial^3 g/\partial x^2 \partial y = \partial^3 g/\partial x \partial y \partial x$?

   (b) In general, does $\partial^3 g/\partial x^2 \partial y = \partial^3 g/\partial x \partial y \partial x$ for any $g(x,y)$? Explain.

6. A careless barber cuts m percent more hair than his customers request according to the following formula: $m = (g + h)/d$, where g is the minutes the barber spends gossipping with the customer; h is the ounces of hair on the customer's head; and d is the expected tip to be received in dollars.

   (a) Compute $\partial m/\partial d$.

   (b) Suppose $\partial m/\partial g = 7$. Explain the meaning of the number.

   (c) What will happen to the customer's hair if the barber expects no tip?

ANSWERS TO PREREQUISITE QUIZ

1. (a) $(\sqrt{x} + 1)[(x^3 + x - 3)/\sqrt{x} + (\sqrt{x} + 1)(3x^2 + 1)]$

   (b) $(2 + \cos x + 2x \sin x)/(2 + \cos x)^3$

   (c) $(\sec^2 x)\exp(\tan x)$

2. 1/300

3. $(d/dx)x^2 = \lim_{\Delta x \to 0} \{[(x + \Delta x)^2 - x^2]/\Delta x\} = \lim_{\Delta x \to 0} [(2x\Delta x + (\Delta x)^2)/\Delta x] = \lim_{\Delta x \to 0} (2x + \Delta x) = 2x$.

4. (i) $\lim_{x \to a} f(x)$ exists and (ii) $\lim_{x \to a} f(x) = f(a)$.

ANSWERS TO SECTION QUIZ

1. x ; y and z .

2. $\partial^2 f/\partial y^2$; compute f_y and differentiate f_y with respect to y .

3. (a) $\lim_{h \to 0} \{[g(x,y,z + h) - g(x,y,z)]/h\}$

    (b) 0

4. $(y^2 e^{3x} - e^x)\sin(e^x y) - 3ye^{2x}\cos(e^x y)$

5. (a) Yes

    (b) Only if the second partial with respect to x is continuous, and if the first partial with respect to y is continuous.

6. (a) $-(g + h)/d^2$

    (b) Holding h and d constant, the barber will cut 7% more hair than requested for every minute he gossips with a customer.

    (c) The barber will cut off all of the customer's hair.

15.2 Linear Approximations and Tangent Planes

PREREQUISITES

1.   Recall how to compute a partial derivative (Section 15.1).

2.   Recall how to find the linear approximation of a function of one
     variable (Section 1.6).

3.   Recall how to find the tangent line to a curve (Section 1.6).

PREREQUISITE QUIZ

1.   Use the linear approximation to estimate the value of $(1.02)^2$.

2.   What is the equation of the tangent line to the curve described by $y =$
     $x^3 + x^2 + 1$ at $x = 2$?

3.   Find $\partial f/\partial x$ and f_y for the following functions:

     (a) $f(x,y) = x + y \cos x$

     (b) $f(x,y) = y - 5$

GOALS

1.   Be able to find the linear approximation for a function of two or three
     variables.

2.   Be able to compute the tangent plane of the graph of functions of two
     or three variables.

STUDY HINTS

1.   <u>Linear approximation</u>. With one variable calculus, we began at $f(x_0)$
     and moved along the tangent line as we made a change in the x-direction
     to obtain an approximation for $f(x)$. Now, we need to make changes in
     other directions than just the x-direction to obtain an approx-
     mation for f . Thus, we get $f(x,y) \approx f(x_0,y_0) + (\partial f/\partial x)\Delta x +$

1. (continued)

   $(\partial f/\partial y)\Delta y$ and $f(x,y,z) \approx f(x_0,y_0,z_0) + (\partial f/\partial x)\Delta x + (\partial f/\partial y)\Delta y +$
   $(\partial f/\partial z)\Delta z$, where the partials need to be evaluated at (x_0,y_0) and
   (x_0,y_0,z_0) , respectively. As with one variable calculus, using a calculator defeats the purpose.

2. Tangent plane. The equation of a tangent plane is the same equation
   which is used to compute the linear approximation. If $f(x,y) = z$,
   the linear approximation equation $z = z_0 + f_x\Delta x + f_y\Delta y$ can be re-
   arranged to read $0 = -f_x\Delta x - f_y\Delta y + \Delta z$. Thus, a normal vector to
   this plane is $-f_x\underline{i} - f_y\underline{j} + \underline{k}$.

SOLUTIONS TO EVERY OTHER ODD EXERCISE

1. The tangent plane is given by $z = f(x_0,y_0) + [f_x(x_0,y_0)](x - x_0) +$
   $[f_y(x_0,y_0)](y - y_0)$. $\partial z/\partial x = 3x^2 - 6y$ and $\partial z/\partial y = 3y^2 - 6x$. At
   $(1,2,-3)$, $\partial z/\partial x = -9$ and $\partial z/\partial y = 6$. Thus, the tangent plane at
   $(1,2,-3)$ is $z = -3 + (-9)(x - 1) + 6(y - 2)$ or $z = -9x + 6y - 6$.

5. Using the tangent plane formula from Exercise 1, we have $f_x = 2x$ and
   $f_y = 6y^2$. This gives $z = 4 + 2(x - 1) + 6(y - 1) = 2x + 6y - 4$.

9. We could apply the method of Exercise 1 with $z_0 = f(x_0,y_0)$. However,
   note that the equation describes a plane, which is its own tangent plane.

13. A normal vector is $-f_x(x_0,y_0)\underline{i} - f_y(x_0,y_0)\underline{j} + \underline{k}$. To normalize it,
    divide by its length, which is $[(f_x(x_0,y_0))^2 + (f_y(x_0,y_0))^2 + 1]^{1/2}$.
    Using the result of Exercise 9, we get $\underline{n} = (-\underline{i} + \underline{j} + \underline{k})/\sqrt{1 + 1 + 1} =$
    $(-1/\sqrt{3})(\underline{i} - \underline{j} - \underline{k})$.

17. Use $f(x,y) \approx f(x_0,y_0) + [f_x(x_0,y_0)](x - x_0) + [f_y(x_0,y_0)](y - y_0)$.
    Let $f(x,y) = x^2(1 - \sqrt{y})$, so $f_x = 2x(1 - \sqrt{y})$ and $f_y = -x^2/(2\sqrt{y})$.
    Also, let $x_0 = 1$, $y_0 = 1.96 = (1.4)^2$, so $f(x,y) \approx f(1,1.96) +$

17. (continued)

$f_x(1,1.96)(x - 1) + f_y(1,1.96)(y - 1.96) = 1(1 - 1.4) + 2(1 - 1.4) \times$

$(x - 1) - 1/(2.8)(y - 1.96) = -0.4 - 0.8(x - 1) - (y - 1.96)/(2.8)$.

Then $(1.01)^2(1 - \sqrt{1.98}) = f(1.01,1.98) \approx -0.4 - 0.8(0.01) - (0.02)/$

$(2.8) \approx -0.408 - 0.007 \approx -0.415$.

21. Use $f(x,y,z) \approx f(x_0,y_0,z_0) + [f_x(x_0,y_0,z_0)](x - x_0) + [f_y(x_0,y_0,z_0)] \times$

$(y - y_0) + [f_z(x_0,y_0,z_0)](z - z_0)$. Let $f(x,y,z) = xyz$, so $f_x = yz$,

$f_y = xz$, and $f_z = xy$. Also, let $x_0 = 1 = y_0 = z_0$, so $f(x,y,z) \approx$

$f(1,1,1) + [f_x(1,1,1)](x - 1) + [f_y(1,1,1)](y - 1) + [f_z(1,1,1)](z - 1)$.

Then $(0.98)(0.99)(1.03) = f(0.98,0.99,1.03) \approx 1 + (-0.02) + (-0.01) +$

$(0.03) = 1.00$.

25. Since a cube has six faces, $f(a,v) = 6v/a$. Therefore, $f_a = -6v/a^2$

and $f_v = 6/a$. Also let $a_0 = 6$ and $v_0 = 1$. According to the

linear approximation, $f(a,v) \approx f(6,1) + f_a(a - 6) + f_v(v - 1) =$

$1 - (a - 6)/6 + (v - 1)$. When $a = 6 + \Delta a$ and $v = 1 + \Delta v$,

$f(6 + \Delta a, 1 + \Delta v) \approx 1 - \Delta a/6 + \Delta v$.

29. Let $\underline{u}$ denote the vector $(x_0,y_0,f(x_0,y_0))$. $f_x = x/(1 - x^2 - y^2)^{1/2}$

and $f_y = y/(1 - x^2 - y^2)^{1/2}$. A normal to the surface and the tangent

plane at (x_0,y_0) is $(-x_0/(1 - x_0^2 - y_0^2)^{1/2}, -y_0/(1 - x_0^2 - y_0^2)^{1/2}, 1)$.

Multiply this by $f(x_0,y_0)$ to produce $\underline{u}$. Hence these vectors are

parallel, so $\underline{u}$ is also normal to the plane.

SECTION QUIZ

1. A box has dimensions $2.015 \times 0.998 \times 1.007$. Use the linear approximation

to estimate its volume.

2.   Estimate the surface area of the box in Question 1.

3.   (a) Find the equation of the tangent plane to the surface $5x^2 + xy^3 - y - z = 0$ at $(1,1,5)$.

   (b) How does the tangent plane in part (a) compare with that to the surface $z = 5x^2 + xy^3 - y$ at $(1,1)$?

   (c) Find a normal unit vector to the plane in part (a).

4.   What is the tangent plane to the unit sphere at $(1/2,-1/2,-1/\sqrt{2})$?

5.   Hillbillies' feuding level (HFL) is known to be determined by HFL = $s^3 t\sqrt{d}$, where s is the number of times family members have seen each others' faces during the day, t is the number of hours since the last attack, and d is the distance (in meters) between the homes.

   (a) Suppose $s = 4$, $t = 3.1$, and $d = 102$. Use the linear approximation to estimate HFL .

   (b) Hillbillies usually start throwing rocks at each other when HFL > 1000 . If $s = 4$ and $t = 3.1$, approximate how close the homes can be without any rock throwing. (Hint: Guess a d_0 for the linear approximation formula and then find a Δd .)

ANSWERS TO PREREQUISITE QUIZ

1.   1.04

2. $y = 16x - 21$

3.   (a) $\partial f/\partial x = 1 - y \sin x$; $f_y = \cos x$.

   (b) $\partial f/\partial x = 0$; $f_y = 1$.

ANSWERS TO SECTION QUIZ

1. 2.025

2. 10.090

3. (a) $11x + 2y - z = 8$

   (b) They are the same.

   (c) $(11\underline{i} + 2\underline{j} - \underline{k})/\sqrt{126}$

4. $x - y - \sqrt{2}z - 2 = 0$

5. (a) 1993.6

   (b) 25.0

15.3 The Chain Rule

PREREQUISITES

1. Recall how to use the chain rule (Section 2.2).

2. Recall the linear approximation formula for two variables (Section 15.2).

PREREQUISITE QUIZ

1. Write an expression for $(d/dx)f(g(x))$.

2. Differentiate $(\sqrt{x} + 1)^2/\exp(\cos x)$.

3. If $f(x,y) = x^2 + xy$, approximate $f(1.1,0.9)$ by using the linear approximation.

GOALS

1. Be able to use the chain rule to compute the partial derivatives of a function which depends on intermediate variables.

2. Be able to use the chain rule to find tangents to curves in graphs.

STUDY HINTS

1. Chain rule. This is a simple extension of the one-variable case. An extra term is added for each variable. Notice that each term has the same variable in the numerator and the denominator. For example, in $du/dt = (\partial u/\partial x)(dx/dt) + (\partial u/\partial y)(dy/dt) + (\partial u/\partial z)(dz/dt)$, there is an x in the numerator and the denominator of the first term. The rule is similar in structure to the chain rule for one variable. These suggestions should make the formula easy to memorize.

2. <u>Tangent lines</u>. If a line is tangent to a curve in a surface, the line
    must lie in the tangent plane. Convince yourself of this fact.

SOLUTIONS TO EVERY OTHER ODD EXERCISE

1. (a) $\partial T/\partial x$ = 2x cos y - y^2cos x ; $\partial T/\partial y$ = $-x^2$sin y - 2y sin x ; dx/dt =

      8 ; and dy/dt = -2 . By the chain rule, dT/dt = $(\partial T/\partial x)$(dx/dt) +

      $(\partial T/\partial y)$(dy/dt) = 16x cos y - $8y^2$cos x + $2x^2$sin y + 4y sin x =

      (48 + 128t)cos(3 - 2t) - 8$(3 - 2t)^2$cos(3 + 8t) +

      2$(3 + 8t)^2$sin(3 - 2t) + (12 - 8t)sin(3 + 8t) .

    (b) Substitution yields T = $(3 + 8t)^2$cos(3 - 2t) - $(3 - 2t)^2$sin(3 + 8t) .

      Therefore, dT/dt = 2(8)(3 + 8t)cos(3 - 2t) + 2$(3 + 8t)^2$sin(3 - 2t) +

      2(2)(3 - 2t)sin(3 + 8t) - $(3 - 2t)^2$(8)cos(3 + 8t) , which is the

      same as in part (a).

5. We find $f'(t)$ in two ways. First we compute f_x , f_y , f_z , dx/dt ,

    dy/dt , and dz/dt and combine them according to the chain rule.

    Secondly, we substitute $\underline{g}(t)$ into the expression for f and differ-

    entiate. The two derivatives will be equal. f_x = 1 ; f_y = 2y ; and

    f_z = $3z^2$. Also, dx/dt = -sin t ; dy/dt = cos t ; and dz/dt = 1 .

    Then the chain rule says $f'(t)$ = -sin t + 2y cos t + $3z^2$.

        Substituting for x , y , and z gives f(t) = cos t + $\sin^2$t +

    t^3 . Therefore $f'(t)$ = -sin t + 2 sin t cos t + $3t^2$. Since y =

    sin t and z = t , -sin t + 2 sin t cos t + $3t^2$ = -sin t + 2y cos t +

    $3z^2$.

9. Let x = f(t) and y = g(t) . Then z = f(t)/g(t) = x/y . Applying

    the chain rule, we have dz/dt = $(\partial z/\partial x)\cdot$(dx/dt) + $(\partial z/\partial y)\cdot$(dy/dt) =

    $(1/y)\cdot f'(t)$ + $(-x/y^2)g'(t)$ = [$yf'(t)$ - $xg'(t)$]$/y^2$ = [g(t)$f'(t)$ -

    f(t)$g'(t)$]$/[g(t)]^2$, which is the quotient rule.

13. Recall that a normal vector to $z = f(x,y)$ is $-f_x(x_0,y_0)\underline{i} - f_y(x_0,y_0)\underline{j} + \underline{k}$. For $z = x^2 + y^2$, $f_x = 2x$ and $f_y = 2y$. Thus, a normal vector at $(1,2,5)$ is $-2\underline{i} - 4\underline{j} + \underline{k}$. All tangent vectors, $a\underline{i} + b\underline{j} + c\underline{k}$, lie in the tangent plane, which is $z = 5 + 2(x - 1) + 4(y - 2)$. The tangent vectors satisfy $-2a - 4b + c = 0$.

17. Let $x = f(t)$, $y = g(t)$, and $z = h(t)$. Then $u = xy/z$ can be differentiated by the chain rule: $du/dt = (\partial u/\partial x)(dx/dt) + (\partial u/\partial y)(dy/dt) + (\partial u/\partial z)(dz/dt) = (y/z)f'(t) + (x/z)g'(t) + (-xy/z^2)h'(t)$. Putting this over a common denominator, z^2 , yields $du/dt = [yzf'(t) + xzg'(t) - xyh'(t)]/z^2 = [f'(t)g(t)h(t) + f(t)g'(t)h(t) - f(t)g(t)h'(t)]/[h(t)]^2$.

21. The half-line is parametrized by (sx_0,sy_0,sz_0) , $s \geq 0$. This curve passes through (x_0,y_0,z_0) at $s = 1$ and by assumption, the entire half-line lies in the surface $z = f(x,y)$. By the box on p. 782, the tangent to this curve lies in the tangent plane. But the tangent at $s = 1$ is the vector (x_0,y_0,z_0) . Thus, (x_0,y_0,z_0) lies in the tangent plane. Parametrizing the half-line by $(s\alpha x_0,s\alpha y_0,s\alpha z_0)$ for fixed $\alpha > 0$ and variable $s \geq 0$, we similarly see that $(\alpha x_0,\alpha y_0,\alpha z_0)$ lies in the tangent plane. Thus the whole half-line lies in it. An example of such a surface is $z = \sqrt{x^2 + y^2}$.

SECTION QUIZ

1. (a) Suppose that P is a function of four variables, w , x , y , and z . If w , x , y , and z are all functions of r , what do you suppose is the chain rule formula for dP/dr ?

1.   (b) Let $P = wz + yx$, $w = 3r$, $x = \cos r$, $y = r^4$, and $z = 1/r$.

Find dP/dr by using the chain rule formula from part (a).

(c) Use substitution to check your answer in part (b).

2.   Suppose that $u = f(x,y)$, $x = g(t)$, and $y = h(t)$. Write a formula
for d^2u/dt^2 . [Hint: Your formula should involve $d(\partial u/\partial x)/dt$.]

3.   The curve described parametrically by $(x,y,z) = (2 \cos t, 3 \sin t, t)$
lies on the elliptical cylinder $9x^2 + 4y^2 = 36$. Show that the tangent
line to the curve lies on the tangent plane to the surface at the point
$(\sqrt{2}, 3\sqrt{2}/2, \pi/4)$.

4.   The flight speed of Santa's reindeer varies according to w , the weight
of the toys they are hauling, and D , the distance they have travelled
on Christmas eve. In addition, it is also known that the weight of the
toys decreases with time and the distance travelled is a function of time.

(a) Let t represent the time elapsed since the reindeer left the North
Pole and let s be their speed at time t . Using the chain rule,
write a formula for ds/dt .

(b) Compute ds/dt in terms of the partials of s if $w = -50t$ and
$D = 75t$.

ANSWERS TO PREREQUISITE QUIZ

1. $f'(g(x)) \cdot g'(x)$

2. $[1 + 1/\sqrt{x} + (\sin x)(\sqrt{x} + 1)^2]/\exp(\cos x)$

3.   2.2

ANSWERS TO SECTION QUIZ

1. (a) $dP/dr = (\partial P/\partial w)(dw/dr) + (\partial P/\partial x)(dx/dr) + (\partial P/\partial y)(dy/dr) +$
      $(\partial P/\partial z)(dz/dr)$

   (b) $4r^3\cos r - r^4\sin r$

   (c) $4r^3\cos r - r^4\sin r$

2. $[d(\partial u/\partial x)/dt](dx/dt) + (\partial u/\partial x)(d^2x/dt^2) + [d(\partial u/\partial y)/dt](dy/dt) +$
   $(\partial u/\partial y)(d^2y/dt^2)$

3. Tangent plane is $18\sqrt{2}x + 12\sqrt{2}y - z = 72 - \pi/4$; tangent line is
   $(-\sqrt{2}, 3\sqrt{2}/2, 1)(t - \pi/4) + (\sqrt{2}, 3\sqrt{2}/2, \pi/4)$. Tangent line satisfies
   tangent plane equation.

4. (a) $ds/dt = (\partial s/\partial w)(dw/dt) + (\partial s/\partial D)(dD/dt)$

   (b) $ds/dt = (\partial s/\partial w)(-50) + (\partial s/\partial D)(75)$

15.4 Matrix Multiplication and the Chain Rule

PREREQUISITES

1. Recall how to find a partial derivative (Section 15.1).

2. Recall how to use the chain rule (Section 15.3).

PREREQUISITE QUIZ

1. Find $\partial f/\partial x$, $\partial f/\partial y$, and $\partial^2 f/\partial x^2$ for the following functions:

    (a) $f(x,y) = x^2 + 3x^2 y - y^2 + y$

    (b) $f(x,y) = e^{x^2 y} - \sin 3y^2$

2. Suppose that f is a function of x and y , and that x and y are
    both functions of t . Write a formula for df/dt .

3. Let $f(x,y) = x^3 + y^2$, $x = e^t$, and $y = t^2$.

    (a) Substitute for x and y , and then compute df/dt .

    (b) Use the chain rule to compute df/dt .

GOALS

1. Be able to multiply matrices.

2. Be able to write down the derivative matrix and use it to obtain the
    general chain rule.

STUDY HINTS

1. Matrices. A matrix is a rectangular array of numbers. An n × m
    matrix has n rows and m columns. You should remember that we
    talk about rows before columns. Thus n × m means n rows and m
    columns. The (i,j) entry is the number located in row i , column j .

2. Matrix multiplication. To obtain the product of matrix A and matrix
    B , we manipulate the entries of row i in A and column j in B to

2.   (continued)

obtain the ij entry of AB . Therefore, the number of columns of A must equal the number of rows of B if AB is to exist. If A is an $n \times m$ matrix and B is $m \times p$, then AB is an $n \times p$ matrix (The m's "cancel").

3.   <u>Notation</u>. $\partial u/\partial(x,y,z)$ is the vector consisting of the partials of u with respect to x , y , and z . $\partial(x,y,z)/\partial t$ is the vector consisting of the partials of x , y , and z with respect to t . In general, similar notation defines a matrix where all of the top variables are differentiated with respect to all of the bottom variables.

4.   <u>Derivative matrix</u>. In the notation $\partial(u_1 , \ldots , u_m)/\partial(x_1 , \ldots , x_n)$, the matrix is written so that the u's are listed down a column. The x's are listed across a row. Another way to help you remember is to realize that an $m \times n$ matrix is required. When you evaluate the matrix at a given point, the point refers to the x's , not the u's .

5.   <u>Non-commutativity</u>. Example 5 shows that $AB \neq BA$ in general. Matrix multiplication is <u>not</u> commutative.

6.   <u>General chain rule</u>. The formula for the general chain rule is $\partial(u_1 , \ldots , u_m)/\partial(t_1 , \ldots , t_k) = [\partial(u_1 , \ldots , u_m)/\partial(x_1 , \ldots , x_n)] \times [\partial(x_1 , \ldots , x_n)/\partial(t_1 , \ldots , t_k)]$, where "$\times$" indicates matrix multiplication. Note how cancellations can be made in this formula as if it was a "fraction." The general formula is the product of two derivative matrices, so any desired partial may be obtained by multiplication.

7.   <u>Second derivatives by chain rule</u>. There are some subtleties which you should be aware of when you compute second derivatives. Suppose that $z = f(x,y)$ and $x = r \cos \theta$, $y = r \sin \theta$. Then $\partial z/\partial \theta =$

7.   (continued)

$(\partial z/\partial x)(\partial x/\partial\theta) + (\partial z/\partial y)(\partial y/\partial\theta)$. Now, $\partial^2 z/\partial\theta^2$ is obtained by using the product and sum rules to get $\partial^2 z/\partial\theta^2 = (\partial^2 z/\partial x\partial\theta)(\partial x/\partial\theta) + (\partial z/\partial x)(\partial^2 x/\partial\theta^2) + (\partial^2 z/\partial y\partial\theta)(\partial y/\partial\theta) + (\partial z/\partial y)(\partial^2 y/\partial\theta^2)$. One of the terms contains $\partial^2 z/\partial x\partial\theta$, and another contains $\partial^2 z/\partial y\partial\theta$, which both involve the expression which we obtained for $\partial z/\partial\theta$. In the second step, many students forget that z depends on θ .

SOLUTIONS TO EVERY OTHER ODD EXERCISE

1. $[1\ 2\ 3]\begin{bmatrix} 4 \\ 5 \\ 6 \end{bmatrix} = [1\cdot 4 + 2\cdot 5 + 3\cdot 6] = [32]$.

5.   Use the method of Example 2.

$\dfrac{\partial(x,y)}{\partial(u,v)} = \begin{bmatrix} \partial x/\partial u & \partial x/\partial v \\ \partial y/\partial u & \partial y/\partial v \end{bmatrix} = \begin{bmatrix} \sin v & u\cos v \\ v\exp(uv) & u\exp(uv) \end{bmatrix}$, which is

$\begin{bmatrix} \sin 1 & 0 \\ 1 & 0 \end{bmatrix}$ at $(0,1)$.

9. $\begin{bmatrix} 1 & 2 \\ 0 & 1 \end{bmatrix}\begin{bmatrix} 2 & 3 \\ 4 & 5 \end{bmatrix} = \begin{bmatrix} 1\cdot 2 + 2\cdot 4 & 1\cdot 3 + 2\cdot 5 \\ 0\cdot 2 + 1\cdot 4 & 0\cdot 3 + 1\cdot 5 \end{bmatrix} = \begin{bmatrix} 10 & 13 \\ 4 & 5 \end{bmatrix}$.

13. $\begin{bmatrix} 0 & 1 \\ 2 & 3 \end{bmatrix}\begin{bmatrix} 1 \\ 2 \\ 3 \end{bmatrix}$ is undefined because the number of columns in the first matrix

(two) is unequal to the number of rows of the second matrix (three).

17. $\begin{bmatrix} 1 & 0 \\ 0 & 0 \end{bmatrix}\begin{bmatrix} a & b \\ c & d \end{bmatrix} = \begin{bmatrix} 1\cdot a + 0\cdot c & 1\cdot b + 0\cdot d \\ 0\cdot a + 0\cdot c & 0\cdot b + 0\cdot d \end{bmatrix} = \begin{bmatrix} a & b \\ 0 & 0 \end{bmatrix}$.

21.   Use the formula $[\partial z/\partial x\ \ \partial z/\partial y] = [\partial z/\partial u\ \ \partial z/\partial v]\begin{bmatrix} \partial u/\partial x & \partial u/\partial y \\ \partial v/\partial x & \partial v/\partial y \end{bmatrix}$. By

matrix multiplication, $[\partial z/\partial x\ \ \partial z/\partial y] = [2u\ \ 2v]\begin{bmatrix} 2 & 0 \\ 3 & 1 \end{bmatrix} = [4u + 6v\ \ \ 2v]$.

In terms of x and y , we have $\partial z/\partial x = 26x + 6y + 70$ and $\partial z/\partial y =$

21. (continued)

$6x + 2y + 14$.

Using direct substitution, $z = (2x + y)^2 + (3x + y + 7)^2$. Thus, $\partial z/\partial x = 2(2x + 7)(2) + 2(3x + y + 7)(3) = 26x + 6y + 70$, and $\partial z/\partial y = 2(3x + y + 7) = 6x + 2y + 14$. The two answers are consistent.

25. (a) $\partial(x,y)/\partial(t,s) = \begin{bmatrix} \partial x/\partial t & \partial x/\partial s \\ \partial y/\partial t & \partial y/\partial x \end{bmatrix} = \begin{bmatrix} 1 & 1 \\ 1 & -1 \end{bmatrix}$. $\partial(u,v)/\partial(x,y) =$

$\begin{bmatrix} \partial u/\partial x & \partial u/\partial y \\ \partial v/\partial x & \partial u/\partial y \end{bmatrix} = \begin{bmatrix} 2x & 2y \\ 2x & -2y \end{bmatrix}$.

(b) $u = (t + s)^2 + (t - s)^2$; $v = (t + s)^2 - (t - s)^2$.

$\partial(u,v)/\partial(t,s) = \begin{bmatrix} \partial u/\partial t & \partial u/\partial s \\ \partial v/\partial t & \partial v/\partial s \end{bmatrix} =$

$\begin{bmatrix} 2(t + s) + 2(t - s) & 2(t + s) - 2(t - s) \\ 2(t + s) - 2(t - s) & 2(t + s) + 2(t - s) \end{bmatrix} = \begin{bmatrix} 4t & 4s \\ 4s & 4t \end{bmatrix}$.

(c) We wish to show that $\partial(u,v)/\partial(t,s) = [\partial(u,v)/\partial(x,y)][\partial(x,y)/\partial(t,s)]$.

The right side is $\begin{bmatrix} 2x & 2y \\ 2x & -2y \end{bmatrix}\begin{bmatrix} 1 & 1 \\ 1 & -1 \end{bmatrix} = \begin{bmatrix} 2x + 2y & 2x - 2y \\ 2x - 2y & 2x + 2y \end{bmatrix}$.

But $2x + 2y = 4t$ and $2x - 2y = 4s$, so the above $= \begin{bmatrix} 4t & 4s \\ 4s & 4t \end{bmatrix} =$

$\partial(u,v)/\partial(t,s)$.

29. By the general chain rule, $[\partial u/\partial r \ \ \partial u/\partial\theta \ \ \partial u/\partial\phi] = [\partial u/\partial x \ \ \partial u/\partial y \ \ \partial u/\partial z] \times$

$\begin{bmatrix} \partial x/\partial r & \partial x/\partial\theta & \partial x/\partial\phi \\ \partial y/\partial r & \partial y/\partial\theta & \partial y/\partial\phi \\ \partial z/\partial r & \partial z/\partial\theta & \partial z/\partial\phi \end{bmatrix} = [\partial u/\partial x \ \ \partial u/\partial y \ \ \partial u/\partial z] \times$

$\begin{bmatrix} \cos\theta\sin\phi & -r\sin\theta\sin\phi & r\cos\theta\cos\phi \\ \sin\theta\sin\phi & r\cos\theta\sin\phi & r\sin\theta\cos\phi \\ \cos\phi & 0 & -r\sin\phi \end{bmatrix}$. Thus, $\partial u/\partial r =$

29. (continued)

cos θ sin φ(∂u/∂x) + sin θ sin φ (∂u/∂y) + cos φ(∂u/∂z) ; ∂u/∂θ =

-r sin θ sin φ(∂u/∂x) + r cos θ sin φ(∂u/∂y) ; and ∂u/∂φ =

r cos θ cos φ(∂u/∂x) + r sin θ cos φ(∂u/∂y) - r sin φ(∂u/∂z) .

33. We have b_i = 1/m for i = 1 , ... , m , so AB = $\sum_{i=1}^{m}$ a_i(1/m) =

(1/m)$\sum_{i=1}^{m}$ a_i . It is just the average of the entries of A .

37. Let B = $\begin{bmatrix} a & b \\ c & d \end{bmatrix}$. Then we want $\begin{bmatrix} a & b \\ c & d \end{bmatrix}\begin{bmatrix} 1 & 2 \\ 2 & 5 \end{bmatrix}$ = $\begin{bmatrix} 1 & 0 \\ 0 & 1 \end{bmatrix}$; that is,

$\begin{bmatrix} a + 2b & 2a + 5b \\ c + 2d & 2c + 5d \end{bmatrix}$ = $\begin{bmatrix} 1 & 0 \\ 0 & 1 \end{bmatrix}$. Hence a + 2b = 1(i) ; 2a + 5b = 0(ii) ;

c + 2d = 0(iii) ; and 2c + 5d = 1(iv) . Subtracting 2 times equa-

tion (i) from equation (ii) , we get b = -2 , so a = 5 . Sub-

tracting 2 times equation (iii) from equation (iv) , we get d =

1 ; so c = -2 . Hence B = $\begin{bmatrix} 5 & -2 \\ -2 & 1 \end{bmatrix}$.

41. (a) $\left| \partial(x,y)/\partial(t,s) \right|_{(1,2)}$ = $\left| \begin{bmatrix} 2t & 2s \\ 2t & -2s \end{bmatrix} \right|_{(1,2)}$ = $\left| \begin{matrix} 2 & 4 \\ 2 & -4 \end{matrix} \right|$ = -16 .

    (b) $\left| \partial(u,v)/\partial(x,y) \right|_{(5,-3)}$ = $\left| \begin{bmatrix} 1 & 1 \\ y & x \end{bmatrix} \right|_{(5,-3)}$ = (x - y)$|_{(5,-3)}$ = 8 .

    (c) We have u = $2t^2$ and v = $t^4 - s^4$, so $\left| \partial(u,v)/\partial(t,s) \right|_{(1,2)}$ =

    $\left| \begin{bmatrix} 4t & 0 \\ 4t^3 & -4s^3 \end{bmatrix} \right|_{(1,2)}$ = $\left| \begin{matrix} 4 & 0 \\ 4 & -32 \end{matrix} \right|$ = 128 , which equals -16·8 .

45. (a) x/au = sin φ cos θ ; y/bu = sin φ sin θ ; and z/cu = cos φ .

    Compute $(x/au)^2 + (y/bu)^2 + (z/cu)^2$ = $sin^2φ \ cos^2θ + sin^2φ \ sin^2θ +$

    $cos^2φ = sin^2φ(cos^2θ + sin^2θ) + cos^2φ = sin^2φ + cos^2φ = 1$. For

45. (a) (continued)

each constant value of u , this describes an ellipsoid centered at the origin.

(b) $x/(au \sin \phi) = \cos \theta$; and $y/(bu \sin \phi) = \sin \theta$. Therefore, $(x/au \sin \phi)^2 + (y/bu \sin \phi)^2 = 1$. But $z/(cu \cos \phi) = 1$, so $(x/(au \sin \phi))^2 + (y/(bu \sin \phi))^2 = (z/(cu \cos \phi))^2$. Therefore, $(x/(a \sin \phi))^2 + (y/(b \sin \phi))^2 = (z/(c \cos \phi))^2$. If ϕ is constant, then for each value of ϕ , the above equation describes an elliptical cone with vertex at the origin.

(c) Substitute $u \sin \phi = x/(a \cos \theta)$ into $y = xb \sin \theta/a \cos \theta$. If θ is constant, this describes a line through the origin in the xy-plane, or a plane in three-dimensional space.

(d) $|\partial(x,y,z)/\partial(u,\phi,\theta)| =$

$$\begin{vmatrix} a \sin \phi \cos \theta & au \cos \phi \cos \theta & -au \sin \phi \sin \theta \\ b \sin \phi \sin \theta & bu \cos \phi \sin \theta & bu \sin \phi \cos \theta \\ c \cos \phi & -cu \sin \phi & 0 \end{vmatrix} =$$

$a \sin \phi \cos \theta (bcu^2 \sin^2 \phi \cos \theta) +$

$au \cos \phi \cos \theta (bcu \sin \phi \cos \phi \cos \theta) -$

$au \sin \phi \sin \theta (-bcu \sin^2 \phi \sin \theta - bcu \cos^2 \phi \sin \theta) =$

$abcu^2 \sin \phi \cos^2 \theta (\sin^2 \phi + \cos^2 \phi) + abcu^2 \sin \phi \sin^2 \theta (\sin^2 \phi + \cos^2 \phi) =$

$abcu^2 \sin \phi (\cos^2 \theta + \sin^2 \theta) = abcu^2 \sin \phi$.

49. Simpson's rule says that $\int_a^b f(x)dx \approx [(b - a)/(3n)] [f(x_0) + 4f(x_1) + 2f(x_2) + 4f(x_3) + 2f(x_4) + \ldots + 2f(x_{n-2}) + 4f(x_{n-1}) + f(x_n)]$. In matrix form, this is

$$\int_a^b f(x)dx \approx [(b - a)/(3n)][f(x_0) \quad f(x_1) \quad \ldots \quad f(x_n)] \begin{bmatrix} 1 \\ 4 \\ 2 \\ \vdots \\ 2 \\ 4 \\ 1 \end{bmatrix} .$$

SECTION QUIZ

1. True or false.

    (a) An $n \times m$ matrix has n columns and m rows.

    (b) An $n \times m$ matrix has n rows and m columns.

    (c) The (i,j) entry is located in row i , column j .

    (d) The (i,j) entry is located in column i , row j .

2. The following matrix sizes for A and B are given. Does AB exist?
    If yes, what is the size of AB ?

    (a) A: $n \times m$; B: $m \times n$

    (b) A: $n \times m$; B: $n \times m$

    (c) A: $n \times m$; B: $m \times p$

3. Write the matrix $\partial(u,v,w)/\partial(x,y)$ for $u(x,y) = x^2 + y^2$, $v(x,y) = -3x^2 y$,
    and $w(x,y) = y^2$.

4. Game show contestants are chosen according to how enthusiastic they
    are, E , and their willingness to perform abnormal acts, P . Now,
    E and P are both functions of greed, g , and insanity level, i .
    Finally, both g and i are functions of wealth, $, the contestant's
    age, a , and the number of children under twenty years of age which
    drive them crazy, c .

    (a) What is $\partial(E,P)/\partial(g,i)$ in matrix form?

    (b) What is $\partial(g,i)/\partial($,a,c)$ in matrix form?

    (c) How is $\partial(E,P)/\partial($,a,c)$ related to your answers in parts (a) and (b)?

ANSWERS TO PREREQUISITE QUIZ

1. (a) $\partial f/\partial x = 2x + 6xy$; $\partial f/\partial y = 3x^2 - 2y + 1$; $\partial^2 f/\partial x^2 = 2 + 6y$

    (b) $\partial f/\partial x = 2xy \exp(x^2 y)$; $\partial f/\partial y = x^2 \exp(x^2 y) - 6y \cos 3y^2$; $\partial^2 f/\partial x^2 =$
         $(2y + 2xy)\exp(x^2 y)$

2. $df/dt = (\partial f/\partial x)(dx/dt) + (\partial f/\partial y)(dy/dt)$

3. (a) $f(x,y) = e^{3t} + t^4$, so $df/dt = 3e^{3t} + 4t^3$.

    (b) $df/dt = (\partial f/\partial x)(dx/dt) + (\partial f/\partial y)(dy/dt) = (3x^2)(e^t) + (2y)(2t) =$
    $3e^{3t} + 4t^3$

ANSWERS TO SECTION QUIZ

1. (a) False

    (b) True

    (c) True

    (d) False

2. (a) Yes; $n \times n$.

    (b) No

    (c) Yes; $n \times p$.

3. $\begin{bmatrix} 2x & 2y \\ -6xy & -3x^2 \\ 0 & 2y \end{bmatrix}$

4. (a) $\begin{bmatrix} \partial E/\partial g & \partial E/\partial i \\ \partial P/\partial g & \partial P/\partial i \end{bmatrix}$

    (b) $\begin{bmatrix} \partial g/\partial \$ & \partial g/\partial a & \partial g/\partial c \\ \partial i/\partial \$ & \partial i/\partial a & \partial i/\partial c \end{bmatrix}$

    (c) Answer in (a) multiplied by answer in (b).

15.R <u>Review Exercises for Chapter 15</u>

SOLUTIONS TO EVERY OTHER ODD EXERCISE

1.   Holding y constant, we get $g_x = \pi \cos(\pi x)/(1 + y^2)$. And holding x

   constant, we get $g_y = -2y \sin(\pi x)/(1 + y^2)^2$.

5.   Holding y and z constant and differentiating gives $h_x = z$. Holding

   x and z constant, we get $h_y = 2y + z$. Holding x and y constant,

   we get $h_z = x + y$.

9.   Recall that $(d/dx)\int_0^x f(t)dt = f(x)$, so $g_x = z + e^z x^2 e^x = z + x^2 e^{x+z}$;

   $g_y = 0$; and $g_z = x + e^z \int_0^x t^2 e^t dt$.

13. $\partial u/\partial x = z^2 + z^3 \sin(xz^3)$, so $\partial^2 u/\partial x \partial z = 2z + 3z^2 \sin(xz^3) + 3xz^5 \cos(xz^3)$.

   On the other hand, $\partial u/\partial z = 2xz + 3xz^2 \sin(xz^3)$, so $\partial^2 u/\partial z \partial x = 2z + $

   $3z^2 \sin(xz^3) + 3xz^5 \cos(xz^3)$, which is also $\partial^2 u/\partial x \partial z$.

17. $(\partial/\partial x)\exp(x - \cos(yx)) = (1 + y \sin(yx))\exp(x - \cos(yx))$. Evaluating

   at (1,0) , we get $\exp(1 - \cos(0)) = e^0 = 1$.

21. (a) Substitute x = 27 and V = 65 into $T = 32V/(x + 32)$ to get

      $(32)(65)/(27 + 32) = 2080/59 \approx 35.25$ minutes.

    (b) (i) $\partial T/\partial x = -32V/(x + 32)^2$. Substitute x = 27 and V = 65 ,

         giving $(\partial T/\partial x)\big|_{(27,65)} = -32 \cdot 65/(27 + 32)^2 = -2080/(59)^2 \approx$

         -0.598 minutes/foot. This means that when you are diving

         at a depth of 27 feet, going a foot deeper decreases the

         possible time of the dive by 0.598 minutes, about 36

         seconds.

      (ii) $\partial T/\partial V = 32/(x + 32)$. Substitute x = 27 , giving

         $(\partial T/\partial V)\big|_{x=27} = 32/59 \approx 0.542$ minutes/cubic foot. This means

         that when you are diving at a depth of 27 feet, bringing an

         extra cubic foot of air along extends the possible time of

         the dive by 0.542 minutes, about 33 seconds.

25. For the limit to exist, we must get the same value no matter how we approach $(0,0)$. Suppose we approach $(0,0)$ along the line $y = rx$, where r is any real number. Then $\lim_{(x,y)\to(0,0)}[(x^3 - y^3)/(x^2 + y^2)] = \lim_{(x,rx)\to(0,0)}[(x^3 - r^3x^3)/(x^2 + r^2x^2)] = [(1 - r^3)/(1 + r^2)]\lim_{x\to 0} x = 0$.

29. The tangent plane is given by $z = f(x_0,y_0) + [f_x(x_0,y_0)](x - x_0) + [f_y(x_0,y_0)](y - y_0)$. Here, $f_x = ye^{xy}$ and $f_y = xe^{xy}$, so at $(0,0)$, $f_x = f_y = 0$ and $f = 1$. Thus, the tangent plane is $z = 1$.

33. The linear approximation for two variables is $\ell(x,y) = f(x_0,y_0) + [f_x(x_0,y_0)](x - x_0) + [f_y(x_0,y_0)](y - y_0)$. Let $f(x,y) = x^y$, so $f_x = yx^{y-1}$ and $f_y = (\ln x)x^y$. Let $(x,y) = (0.999,1.001)$ and $(x_0,y_0) = (1,1)$. Then the linear approximation is $1 + 1(-0.001) + 0 = 0.999$.

37. Rewriting $x^2 + 2y^2 + 3z^2 = 6$, we get $z = \sqrt{-x^2/3 - 2y^2/3 + 2}$. Thus, $f_x = -x/3z$ and $f_y = -2y/3z$. The normal is $-f_x\underline{i} - f_y\underline{j} + \underline{k}$. At $(1,1,1)$, the normal is $(1/3)\underline{i} + (2/3)\underline{j} + \underline{k}$. The particle travels along the line $x = 1 + t$, $y = 1 + 2t$, $z = 1 + 3t$, which intersects the surface $x^2 + y^2 + z^2 = 103$ at the point P when $(1 + t)^2 + (1 + 2t)^2 + (1 + 3t)^2 = 103$. Solve this: $3 + 12t + 14t^2 = 103$, so $7t^2 + 6t - 100 = 0$, giving $t = (-6 \pm \sqrt{36 + 2800})/14 = (-3 \pm 2\sqrt{709})/7$. Since $t > 0$, we take $t = (-3 + 2\sqrt{709})/7$. The distance from $(1,1,1)$ to P is $\sqrt{(x - 1)^2 + (y - 1)^2 + (z - 1)^2} = \sqrt{t^2 + 4t^2 + 9t^2} = t\sqrt{14} = \sqrt{14}(-3 + 2\sqrt{709})/7$. Hence, the time is $\sqrt{14}(-3 + 2\sqrt{709})/70$.

41. A function is constant if the derivative is 0. Differentiate $u(f(t),t) = u(x,t)$ by the chain rule to get $(\partial u/\partial x)(dx/dt) + (\partial u/\partial t)(dt/dt) = u_x(dx/dt) + u_t$. Since $dx/dt = u$, we get $uu_x + u_t$, which is given to be 0; therefore, $u(f(t),t)$ is constant in t.

45. Let $x = f(t)$, $y = g(t)$, and $u = \exp(xy)$. Then $du/dt =$
$(\partial u/\partial x)(dx/dt) + (\partial u/\partial y)(dy/dt) = (y \exp(xy))(f'(t)) + (x \exp(xy)) \times$
$(g'(t)) = [f'(t)g(t) + f(t)g'(t)]\exp[f(t)g(t)]$.

49. The tangent plane is given by $z = f(1,1) + f_x(1,1)(x - 1) +$
$f_y(1,1)(y - 1)$. $f_x = 2x$ and $f_y = 12y$, so $z = 7 + 2(x - 1) +$
$12(y - 1) = 2x + 12y - 7$. In the xy-plane, $z = 0$, so the line is
$2x + 12y = 7$, i.e., $y = -x/6 + 7/12$.

53. $\begin{bmatrix} 0 & -1 \\ 1 & 0 \end{bmatrix}\begin{bmatrix} 1 & 0 \\ 0 & -1 \end{bmatrix} = \begin{bmatrix} 0 \cdot 1 - 1 \cdot 0 & 0 \cdot 0 + (-1)(-1) \\ 1 \cdot 1 + 0 \cdot 0 & 1 \cdot 0 + 0(-1) \end{bmatrix} = \begin{bmatrix} 0 & 1 \\ 1 & 0 \end{bmatrix}$.

57. $\begin{bmatrix} 1 & 2 \\ -1 & 1 \\ 2 & 1 \end{bmatrix}\begin{bmatrix} 2 & 1 & 2 \\ -1 & -1 & 1 \end{bmatrix} = \begin{bmatrix} 1(2) + 2(-1) & 1(1) + 2(-1) & 1(2) + 2(1) \\ -1(2) + 1(-1) & -1(1) + 1(-1) & -1(2) + 1(1) \\ 2(2) + 1(-1) & 2(1) + 1(-1) & 2(2) + 1(1) \end{bmatrix} =$

$\begin{bmatrix} 0 & -1 & 4 \\ -3 & -2 & -1 \\ 3 & 1 & 5 \end{bmatrix}$.

61. (a) $z = (u^2 + v^2)/(u^2 - v^2) = (e^{-2x-2y} + e^{2xy})/(e^{-2x-2y} - e^{2xy})$, so
$\partial z/\partial x = [(-2e^{-2x-2y} + 2ye^{2xy})(e^{-2x-2y} - e^{2xy}) - (e^{-2x-2y} + e^{2xy}) \times$
$(-2e^{-2x-2y} - 2ye^{2xy})]/(e^{-2x-2y} - e^{2xy})^2 = 4(e^{-2x-2y+2xy})(1 + y)/$
$(e^{-2x-2y} - e^{2xy})^2$. $\partial z/\partial y = 4(e^{-2y-2x+2xy})(1 + x)/(e^{-2x-2y} - e^{2xy})^2$
because of the symmetry of x and y in u and v .

    (b) By the chain rule, $\partial z/\partial x = (\partial z/\partial u)(\partial u/\partial x) + (\partial z/\partial v)(\partial v/\partial x) =$
$[-4uv^2/(u^2 - v^2)^2](-e^{-x-y}) + [4u^2v/(u^2 - v^2)^2](ye^{xy}) = (4u^2v^2 +$
$4yu^2v^2)/(u^2 - v^2)^2 = 4(e^{-2x-2y+2xy})(1 + y)/(e^{-2x-2y} - e^{2xy})^2$.
Similarly, $\partial z/\partial y = (\partial z/\partial u)(\partial u/\partial y) + (\partial z/\partial v)(\partial v/\partial y)$. Again, the
symmetry of x and y in u and v gives $\partial z/\partial y = 4(e^{-2x-2y+2xy}) \times$
$(1 + x)/(e^{-2x-2y} - e^{2xy})^2$.

65. (a) Merely divide both sides of $PV = nRT$ by the appropriate factor
to get: $n = PV/RT$; $P = nRT/V$; $T = PV/nR$; $V = nRT/P$.

65. (b) $\partial P/\partial T$ is the increase in pressure relative to the increase in

(absolute) temperature, when the number of moles and the volume

are held constant. Differentiate $P = nRT/V$ with respect to T:

$\partial P/\partial T = nR/V$. For example, given two moles of a gas at $300°K$,

a constant volume of 4 liters, and a pressure of 150R Newtons/

meter^2 , the rate of change of pressure with temperature would be

$2R/4 = R/2$. That is, a $1°$ change in temperature would produce

a change of about $R/2$ (N/m^2) in pressure.

(c) Differentiate the appropriate equation from part (a): $\partial V/\partial T =$

nR/P ; $\partial T/\partial P = V/nR$; $\partial P/\partial V = -nRT/V^2$. Multiply to get

$\partial V/\partial T \cdot \partial T/\partial P \cdot \partial P/\partial V = nR/P \cdot V/nR \cdot (-nRT)/V^2 = -nRT/PV = -1$, since

$PV = nRT$.

69. Here, we are differentiating with respect to x to find $\partial w/\partial x$. At

the same time, we hold y constant. However, $y = x^2$, and if y is

constant, x can not be variable.

73. (a) $f_x = 2x/(x^2 + y^2)$, so $f_{xx} = [2(x^2 + y^2) - 2x(2x)]/(x^2 + y^2)^2 =$

$(2y^2 - 2x^2)/(x^2 + y^2)^2$. Similarly, the symmetry of the function

gives $f_{yy} = (2x^2 - 2y^2)/(x^2 + y^2)^2$. Therefore, $f_{xx} + f_{yy} = 0$.

(b) $g_x = (-1/2)(2x)/(x^2 + y^2 + z^2)^{3/2} = -x/(x^2 + y^2 + z^2)^{3/2}$, so

$g_{xx} = [-(x^2 + y^2 + z^2)^{3/2} + x(3/2)(x^2 + y^2 + z^2)^{1/2}(2x)]/$

$(x^2 + y^2 + z^2)^3 = (2x^2 - y^2 - z^2)/(x^2 + y^2 + z^2)^{5/2}$. By symmetry,

$g_{yy} = (2y^2 - x^2 - z^2)/(x^2 + y^2 + z^2)^{5/2}$ and $g_{zz} = (2z^2 - x^2 - y^2)/$

$(x^2 + y^2 + z^2)^{5/2}$. Therefore, $g_{xx} + g_{yy} + g_{zz} = 0$.

(c) $h_x = -2x/(x^2 + y^2 + z^2 + w^2)^2$, so $h_{xx} = [-2(x^2 + y^2 + z^2 + w^2)^2 +$

$2x(2)(x^2 + y^2 + z^2 + w^2)(2x)]/(x^2 + y^2 + z^2 + w^2)^4 = (6x^2 - 2y^2 -$

$2z^2 - 2w^2)/(x^2 + y^2 + z^2 + w^2)^3$. Similarly, symmetry yields $h_{yy} =$

$(6y^2 - 2x^2 - 2z^2 - 2w^2)/(x^2 + y^2 + z^2 + w^2)^3$, $h_{zz} = (6z^2 - 2x^2 -$

73. (c) (continued)

$2y^2 - 2w^2)/(x^2 + y^2 + z^2 + w^2)^3$, and $h_{ww} = (6w^2 - 2x^2 - 2y^2 - 2z^2)/(x^2 + y^2 + z^2 + w^2)^3$. Therefore, $h_{xx} + h_{yy} + h_{zz} + h_{ww} = 0$.

77. $\partial^2 z/\partial x \partial y$ measures the rate of change of $\partial z/\partial y$, the slope of the graph in the y-direction, as x varies. From Fig. 15.R.1, we see that $\partial z/\partial y$ is increasing as x is increased, moving along the line y = 0 . Thus, it is plausible that $(\partial^2 z/\partial x \partial y)|_{(0,0)} > 0$. In fact, we calculate $(\partial z/\partial y)|_{y=0} = x$ and so $(\partial^2 z/\partial x \partial y)|_{(0,0)} = 1$. Also from the graph, $\partial z/\partial x$ is decreasing as y is increased, moving along the line x = 0 . Thus it is plausible that $(\partial^2 z/\partial y \partial x)|_{(0,0)} < 0$. In fact, $(\partial^2 z/\partial y \partial x)|_{(0,0)} = -1$ as above. Thus, the inequality of the mixed partials is consistent with the graph.

TEST FOR CHAPTER 15

1. True or false.

(a) $\partial^2 f/\partial x \partial y = \partial^2 f/\partial y \partial x$ if and only if $\partial f/\partial x = \partial f/\partial y$ at (x_0, y_0) .

(b) If a curve lies on a surface, the tangent line to the curve at (x_0, y_0, z_0) always lies on the tangent plane to the surface at (x_0, y_0, z_0) .

(c) $\partial(x,y,z)/\partial(r,s)$ can be represented by a matrix composed of 3 rows and 2 columns.

(d) If $\lim_{(x,y)\to(0,0)} f(x,y) = 0$, then $\lim_{x\to0} f(x,0) = 0$.

(e) If $\lim_{x\to0} f(x,0) = 0$ and $\lim_{y\to0} f(0,y) = 0$, then $\lim_{(x,y)\to(0,0)} f(x,y) = 0$

2. (a) Sketch the graph of $z = |y|$ in space.

(b) Does $\partial z/\partial x$ exist at the origin? If yes, what is it?

(c) Does $\partial z/\partial y$ exist at the origin? If yes, what is it?

3.   (a) Compute $\partial^3 f/\partial x^2 \partial y$ and $\partial^3 f/\partial x \partial y^2$ for $f(x,y) = 5x^3 + 3y^2$.

   (b) Does the equality of mixed partials apply to the question in part (a)?
        Why or why not?

4.   Recall that in spherical coordinates, $y = \rho \sin \phi \sin \theta$. Suppose that
     ρ , ϕ , and θ are functions of time, t . Find a formula for
     $d^2 y/dt^2$ in terms of the first and second derivatives of ρ , ϕ , and θ .

5.   (a) If $u = f(r,s,p)$, write $\partial u/\partial s$ as a limit.

   (b) For a function, $f(x,y)$, define what it means for f to be
        continuous at (x_0, y_0) .

6.   Let $A = \begin{bmatrix} 1 & 4 \\ 2 & 1 \\ 3 & 2 \end{bmatrix}$ and let $B = \begin{bmatrix} 1 & 2 & 3 \\ 4 & 1 & 2 \end{bmatrix}$. Compute AB and BA if

     they exist. If either does not exist, explain why.

7.   Let $k = rm^2/p$.

   (a) Use the linear approximation to estimate k when $r = 2.1$,
        $m = 1.95$, and $p = 5.2$.

   (b) What is the tangent plane to k at $(r,m,p) = (5,1,2)$?

8.   Let $z = f(x,y) = (x^2 - y^2)/(x^2 + y^2)$.

   (a) Find the vector which is normal to the surface at $(1,1,0)$.

   (b) Prove that $\lim\limits_{(x,y) \to (0,0)} f(x,y)$ does not exist.

9.   (a) Compute $\partial t/\partial r$ if $r = st$.

   (b) Compute $D_3 f$ if $f(x,y,z) = \ln \sqrt{xyz}$.

   (c) Compute h_b if $h(b,k,g) = bk \sin(gb)$.

10. Teenagers are known to break out with zits according to the following
     equation: $z = (f^3 + cf)/w$, where z is the number of new pimples
     breaking out each month, f is the number of pounds of fatty foods
     eaten in the previous month, c is the number of ounces of chocolate
     eaten in the previous month, and w is the number of hours spent
     washing the face in the previous month.

     (a) Find $\partial z/\partial f$ when $f = 10$, $c = 50$, and $w = 3$.

     (b) What is the physical interpretation of $\partial z/\partial f$?

     (c) Suppose f , c , and w are functions of h , the state of
          hunger. (w is a function of h because feeding oneself consumes
          washing time.) Write an expression for dz/dh .

     (d) What would happen to a teenager's face if $w = 0$?

ANSWERS TO CHAPTER TEST

1.   (a) False; only needs continuous second partial derivatives.

     (b) True

     (c) False; matrix is 3 columns by 2 rows.

     (d) True

     (e) False; let $f(x,y) = \begin{cases} 0 & \text{if } x = 0 \text{ or } y = 0 \\ 1 & \text{if } x \neq 0 \text{ or } y \neq 0 \end{cases}$.

2.   (a)

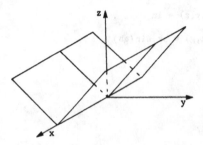

2. (b) Yes; 0 .

    (c) No

3. (a) Both are 0 .

    (b) No; the theorem applies when the partials are taken with respect

         to the same variables the same number of times.

4. $d^2y/dt^2 = [\sin\phi\cos\theta(d\theta/dt) + \cos\phi\sin\theta(d\phi/dt)](d\rho/dt) +$

    $[\sin\theta\cos\phi(d\rho/dt) + \rho\cos\theta\cos\phi(d\theta/dt) - \rho\sin\theta\sin\phi(d\phi/dt)](d\phi/dt) +$

    $[\sin\phi\cos\theta(d\rho/dt) + \rho\cos\phi\cos\theta(d\phi/dt) - \rho\sin\phi\sin\theta(d\theta/dt)](d\theta/dt) +$

    $\sin\phi\sin\theta(d^2\rho/dt^2) + \rho\sin\theta\cos\phi(d^2\phi/dt^2) + \rho\sin\phi\cos\theta(d^2\theta/dt^2)$

5. (a) $\lim\limits_{\Delta s\to 0}\{[f(r,s+\Delta s,p) - f(r,s,p)]/\Delta s\}$

    (b) If f is defined in a disk about (x_0,y_0) , then f is continuous

         at (x_0,y_0) provided $\lim\limits_{(x,y)\to(x_0,y_0)} f(x,y) = f(x_0,y_0)$.

6. $AB = \begin{bmatrix} 17 & 6 & 11 \\ 6 & 5 & 8 \\ 11 & 8 & 13 \end{bmatrix}$; $BA = \begin{bmatrix} 14 & 12 \\ 12 & 21 \end{bmatrix}$.

7. (a) $192/125 = 1.536$

    (b) $k = 2.5 + (r-5)/2 + 5(m-1) - 5(p-2)/4$

8. (a) $\underline{i} - \underline{j} - \underline{k}$

    (b) $\lim\limits_{y\to 0} f(0,y) = -1$ and $\lim\limits_{x\to 0} f(x,0) = +1$.

9. (a) $1/s$

    (b) $1/2z$

    (c) $k\sin(gb) + gbk\cos(gb)$

10. (a) $350/3$

    (b) The number of new pimples for each pound of fatty foods eaten in the

         previous month.

    (c) $dz/dh = (\partial z/\partial f)(df/dh) + (\partial z/\partial c)(dc/dh) + (\partial z/\partial w)(dw/dh)$

    (d) There would be infinitely many pimples on the face.

COMPREHENSIVE TEST FOR CHAPTERS 13 - 15 (Time limit: 3 hours)

1. True or false. If false, explain why.

(a) The derivative of a vector function is <u>always</u> another vector function.

(b) For matrices A and B , AB never equals BA .

(c) If x , y , and z are parametrized by linear functions, the curve in space described by x , y , and z is a straight line.

(d) The value of a 3×3 determinant may not equal the value of a 2×2 determinant.

(e) The dot product of a vector in the plane and a vector in space does not exist, i.e., $(3\underline{i} + \underline{j}) \cdot (\underline{i} + \underline{j} + \underline{k})$ does not exist.

(f) $x^2 + y^2 + z^2 = 1$ for $z \geqslant 0$ is a function of x and y .

(g) The unit hemisphere for $x \geqslant 0$ is the graph of a function of x and y .

(h) If $f(x,y) = 3x^2 + y^2$, then all level curves, except for $c = 0$, are ellipses.

(i) All equations of the form $f(x,y) = ay + b$ represent straight lines in space.

(j) If $\underline{u}$ and $\underline{v}$ are vector functions of t , then $(d/dt)(\underline{u}/\underline{v}) = (\underline{u}' \cdot \underline{v} - \underline{u} \cdot \underline{v}')/(\underline{v} \cdot \underline{v})$.

2. Multiple choice.

(a) If $g(p,s)$ is a function of two variables, then $\partial^2 g / \partial p \partial s = \partial^2 g / \partial s \partial p$ unless:

(i) $\partial g / \partial s = \partial g / \partial p = 0$ at (s_0, p_0) .

(ii) the third partial derivatives are not continuous.

(iii) the second partial derivatives are not continuous.

(iv) none of the above; it is always true.

2. (b) The spherical coordinates of a point are $(4,-3\pi/4,2\pi/3)$. The
        corresponding cylindrical coordinates are:

    (i) $(-\sqrt{12},3\pi/4,-2)$.

    (ii) $(-\sqrt{6},-\sqrt{6},-2)$.

    (iii) $(\sqrt{12},-3\pi/4,-2)$.

    (iv) none of the above.

  (c) The graph of $2x^2 + 5xy + y^2 = 0$ is:

    (i)   a translated circle.

    (ii) a rotated ellipse.

    (iii) a rotated hyperbola.

    (iv) a translated ellipse.

  (d) The equation of the line containing $(-1,0,3)$ and pointing in
       the direction of $2\underline{i} - \underline{j} + \underline{k}$ is:

    (i) $(x,y,z) = (-1,0,3) + (2,-1,1)t$.

    (ii) $(x +1)/2 = -y = z - 3 = t$.

    (iii) both of the above.

    (iv) none of the above.

  (e) If $f(x,y) = \exp(xy) + 3$, then $\partial f/\partial x$ is:

    (i) $x \exp(xy)$.

    (ii) $y \exp(xy)$.

    (iii) $x \exp(xy) + y \exp(xy)$

    (iv) $ye^{xy}\underline{i} + xe^{xy}\underline{j}$.

3. Differentiation problems.

  (a) If $\underline{v}(t) = (t^2 - t)\underline{i} + [(5 - t)/(t^2 - 1)]\underline{j} + t\underline{k}$, what is
       $(d/dp)\underline{v}(pe^{-p})$?

  (b) If $\underline{u}(t) = (t^3 - t^2 + 1)\underline{i} - (5 \ln t)\underline{j}$ and $\underline{v}(t) = 3te^t\underline{i} -$
       $(6 \cos t - 4t)\underline{j} + (5 \sin(t/2))\underline{k}$, what is $(d/dt)[\underline{u}(t) \times \underline{v}(t)]$?
       Do not simplify.

3.   (c) Let $x = u^2 - v^2$, $y = u^2 - 3uv$, $u = re^s$, and $v = \sin(rs)$.

Write $\partial(x,y)/\partial(r,s)$ as a product of matrices.

4.   A surface has the equation $x^2 + 4y^2 - z^2 = 4$.

(a) Describe the level curves for $z = C$, a constant.

(b) Sketch the surface.

(c) At the point $(1,1,1)$, is z changing more rapidly as x

changes or more rapidly as y changes? Explain.

5.   Let $A = (3,5,7)$, $B = (-1,0,-1)$, and $C = (0,6,2)$ be three points
in space.

(a) Find the area of the triangle with vertices at A , B , and C .

(b) For the triangle in part (a), what is $\cos(\angle B)$?

(c) Find the equation of the plane containing A , B , and C .

(d) Compute a unit normal vector to the plane in part (c).

6.   At the carnival, a bean bag is thrown at a human target. Its position
vector is $-20t\underline{i}$; however, fans are displacing the bean bag's
position by the vector $3\underline{i} + t\underline{j} - t\underline{k}$.

(a) What would be the bean bag's position vector if the fans stopped
blowing?

(b) How far would the bean bag travel from $t = 0$ to $t = 1$ if the
fans had stopped blowing?

(c) What is the curvature of the bean bag's path in still air at $t = 1$?

7.
Let $A = \begin{bmatrix} 1 & 2 & 3 \\ 2 & 1 & -1 \\ 3 & -1 & 0 \end{bmatrix}$.

(a) Compute $\det[A]$.

(b) Interpret your answer in part (a) geometrically. Your answer should
involve all of the entries of A .

(c) If a 3×3 determinant is zero, what can you say about the geometry?

8. Let $\underline{u} = -\underline{i} - \underline{j}$ and $\underline{v} = \underline{i} + \underline{j} + \underline{k}$.

(a) Sketch $2\underline{u}$, $\underline{v}$, and $2\underline{u} - \underline{v}$ on the same graph.

(b) Compute the orthogonal projection of $\underline{u}$ on $2\underline{v}$.

(c) Find a unit vector $\underline{p}$, which is orthogonal to both $\underline{u}$ and $\underline{v}$, and such that $\underline{u}$, $\underline{v}$, $\underline{p}$ form a left-handed system.

9. Miscellaneous questions.

(a) For 2×2 matrices A and B , show that $\det(AB) = (\det A) \times (\det B)$.

(b) Find the distance from the point $(1,-1,2)$ to the line passing through $(5,-3,0)$ and $(7,6,-1)$.

(c) Use the linear approximation to estimate $f(x,y,z) = xy + 2xz + 3yz$ at the point $(-1,0.97,2.05)$.

10. Practical application.

A drunk driver has to serve three days in jail. The jailbird consumes q_{b1} slices of bread on day 1, q_{b2} slices on day 2, and q_{b3} slices on day 3. He drinks q_{w1} cups of water on day 1, q_{w2} on day 2, and q_{w3} on day 3. The price of bread per slice is p_b . The price of water per cup is p_w .

(a) What does each entry of the product $[p_b \;\; p_w] \begin{bmatrix} q_{b1} & q_{b2} & q_{b3} \\ q_{w1} & q_{w2} & q_{w3} \end{bmatrix}$ tell you?

(b) If the product in part (a) is AB and you wanted to compute ABC , would you choose C to be $\begin{bmatrix} 1 \\ 1 \\ 1 \end{bmatrix}$ or $[1 \;\; 1 \;\; 1]$?

(c) By performing the multiplication described in part (b), what does your answer tell you?

ANSWERS TO COMPREHENSIVE TEST

1. (a) True

    (b) False; let $A = B = \begin{bmatrix} 1 & 0 \\ 0 & 1 \end{bmatrix}$.

    (c) True

    (d) False; consider $\begin{vmatrix} 1 & 0 & 0 \\ 0 & 1 & 0 \\ 0 & 0 & 1 \end{vmatrix}$ and $\begin{vmatrix} 1 & 0 \\ 0 & 1 \end{vmatrix}$.

    (e) False; the vector $a\underline{i} + b\underline{j}$ is the same as $a\underline{i} + b\underline{j} + 0\underline{k}$.

    (f) True

    (g) False; for every (x,y) in the domain, there are two corresponding

         z values except on the xy-plane.

    (h) True

    (i) False; they are planes.

    (j) False; division of vectors is undefined.

2. (a) iii

    (b) iii

    (c) iii

    (d) iii

    (e) ii

3. (a) $(1 - p)e^{-p}\{(2pe^{-p} - 1)\underline{i} + [(p^2e^{-2p} - 10pe^{-p} + 1)/(p^2e^{-2p} - 1)^2]\underline{j} + \underline{k}\}$

    (b) $[(3t^2 - 2t)\underline{i} - (5/t)\underline{j}] \times [3te^t\underline{i} - (6\cos t - 4t)\underline{j} + (5\sin(t/2))\underline{k}] +$

         $[(t^3 - t^2 + 1)\underline{i} - (5\ln t)\underline{j}] \times [(3 + 3t)e^t\underline{i} - (6\sin t - 4)\underline{j} +$

         $((5/2)\cos(t/2))\underline{k}]$

    (c) $\begin{bmatrix} 2u & -2v \\ 2u - 3v & -3u \end{bmatrix} \begin{bmatrix} e^s & re^s \\ s\cos(rs) & r\cos(rs) \end{bmatrix}$

4.   (a)   Ellipses with minor axis parallel to the y-axis having length
$\sqrt{4 + c^2}$ and with major axis parallel to the x-axis having
length $2\sqrt{4 + c^2}$.

(b)

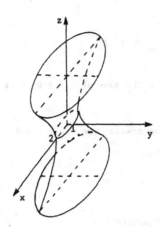

(c)   As y changes; because at $(1,1,1)$, $\partial z/\partial x = 1$ and $\partial z/\partial y = 4$.

5.   (a) $\sqrt{1466}/2$

(b) $58/\sqrt{4830}$

(c) $-33x - 4y + 19z - 14 = 0$

(d) $(-33\underline{i} - 4\underline{j} + 19\underline{k})/\sqrt{1466}$

6.   (a) $(-20t - 3)\underline{i} - t\underline{j} + t\underline{k}$

(b) $\sqrt{402}$

(c) $\sqrt{2}/531\sqrt{59}$

7.   (a) -22

(b)   The volume of the parallelepiped spanned by the vectors $\underline{i} + 2\underline{j} + 3\underline{k}$, $2\underline{i} + \underline{j} - \underline{k}$, and $3\underline{i} - \underline{j}$ is 22 .

(c)   The three vectors lie in a plane or on a line.

8.   (a)

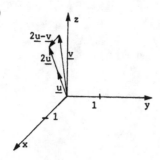

(b) $(2/3)(\underline{i} + \underline{j} + \underline{k})$

(c) $(\underline{i} - \underline{j})/\sqrt{2}$

9.   (a) Let $A = \begin{bmatrix} a & b \\ c & d \end{bmatrix}$ and $B = \begin{bmatrix} e & f \\ g & h \end{bmatrix}$. Both sides equal adeh +

bcgf - adgf - bceh .

(b) $\sqrt{1000/43}$

(c) 0.90

10.   (a) How much it costs to feed bread and water to the prisoner each day.

(b) $\begin{bmatrix} 1 \\ 1 \\ 1 \end{bmatrix}$

(c) The total cost of feeding bread and water to the prisoner during

his prison term.

16.1 Gradients and Directional Derivatives

PREREQUISITES

1. Recall how to use the chain rule for several variables (Section 15.3).

2. Recall how to compute a dot product (Section 13.4).

3. Recall how to draw vectors (Section 13.1).

PREREQUISITE QUIZ

1. Compute the following dot products:

   (a) $(\underline{i} + 2\underline{j} - \underline{k}) \cdot (3\underline{i} - \underline{j} + \underline{k})$

   (b) $(\underline{i} - 3\underline{j}) \cdot (2\underline{j} + 3\underline{k})$

2. If $\underline{u}$ and $\underline{v}$ are non-zero vectors and $\underline{u} \cdot \underline{v} = 0$, how is $\underline{u}$ geometrically related to $\underline{v}$?

3. Suppose that $f(x,y,z) = xy \sin z$ and $(x,y,z) = (t,t^3,t^2)$. Use the chain rule to find df/dt .

4. If P is the point $(1,2)$, draw the vector $\overrightarrow{PQ} = \underline{i} - \underline{j}$.

GOALS

1. Be able to compute the gradient of a given function.

2. Be able to compute a directional derivative and understand its meaning.

3. Be able to determine the direction of the most rapid or least rapid rate of change.

STUDY HINTS

1.   Vector field. This is a function which describes a set of vectors which depend upon their location in the plane or in space. Graphically, we draw these vectors with their tails at the given point in the domain. See Fig. 16.1.1.

2.   Gradients. The gradient of a function is a vector field. The gradient operation is denoted by the symbol $\underline{\nabla}$. Sometimes, it is denoted grad. Thus, $\underline{\nabla}f$ = grad f is the gradient of f . The components of the gradient are simply the partial derivatives of the given function, so $\underline{\nabla}f = (\partial f/\partial x)\underline{i} + (\partial f/\partial y)\underline{j} + (\partial f/\partial z)\underline{k}$. Memorize the formula for the gradient.

3.   Chain rule. The first paragraph of Section 16.1 explains how the chain rule may be thought of as a dot product. It is the gradient of the function dotted with a velocity vector. Thus, $(d/dt)f(\underline{\sigma}(t)) = \underline{\nabla}f(\underline{\sigma}(t))\cdot\underline{\sigma}'(t)$.

4.   Directional derivatives. This concept tells you that a rate of change depends on the direction of movement. For example, if you were climbing a hill, the rate of change of height is greatest going straight up rather than sideways. The computational formula is $\underline{\nabla}f(\underline{r})\cdot\underline{d}$, where $\underline{d}$ is a unit vector. Memorize this formula. Forgetting that $\underline{d}$ is a unit vector is a common mistake. Finally, the directional derivative is a scalar, not a vector.

5.   Gradient interpretation. Since the directional derivative is $\underline{\nabla}f(\underline{r})\cdot\underline{d} = \|\underline{\nabla}f(\underline{r})\|\cdot\|\underline{d}\|\cos\theta = \|\underline{\nabla}f(\underline{r})\|\cos\theta$, the maximum rate of change occurs when $\cos\theta$ is maximized. This occurs when $\theta = 0$. Thus, the greatest increase occurs in the direction of $\underline{\nabla}f$ and the greatest decrease occurs in the direction of $-\underline{\nabla}f$.

6. **Partial derivatives.** Notice that the directional derivative of f in the direction of $\underline{i}$, $\underline{j}$, or $\underline{k}$ is simply the corresponding partial derivative, f_x , f_y , or f_z .

SOLUTIONS TO EVERY OTHER ODD EXERCISE

1. The gradient of $\underline{\nabla}f(x,y,z) = f_x\underline{i} + f_y\underline{j} + f_z\underline{k}$. $f_x(x,y,z) = x/\sqrt{x^2 + y^2 + z^2}$, $f_y(x,y,z) = y/\sqrt{x^2 + y^2 + z^2}$, and $f_z(x,y,z) = z/\sqrt{x^2 + y^2 + z^2}$. Thus, $\underline{\nabla}f(x,y,z) = (x/\sqrt{x^2 + y^2 + z^2})\underline{i} + (y/\sqrt{x^2 + y^2 + z^2})\underline{j} + (z/\sqrt{x^2 + y^2 + z^2})\underline{k}$.

5. The gradient is $\underline{\nabla}f(x,y) = f_x\underline{i} + f_y\underline{j}$. $\ln\sqrt{x^2 + y^2}$ simplifies to $(1/2)\ln(x^2 + y^2)$, so $f_x = x/(x^2 + y^2)$ and $f_y = y/(x^2 + y^2)$. Thus, $\underline{\nabla}f(x,y) = [x/(x^2 + y^2)]\underline{i} + [y/(x^2 + y^2)]\underline{j}$.

9.

$f_x = x/4$ and $f_y = y/6$, so the gradient vector field is $(x/4)\underline{i} + (y/6)\underline{j}$. At each point (x_0,y_0) , we sketch $\underline{\nabla}f(x_0,y_0)$.

13. $r = \sqrt{x^2 + y^2 + z^2}$, so $(\partial/\partial x)(1/(x^2 + y^2 + z^2)) = -2x/(x^2 + y^2 + z^2)^2$. Similarly, $(\partial/\partial y)(1/(x^2 + y^2 + z^2)) = -2y/(x^2 + y^2 + z^2)^2$ and $(\partial/\partial z)(1/(x^2 + y^2 + z^2)) = -2z/(x^2 + y^2 + z^2)^2$. Thus, $\underline{\nabla}(1/r^2) = -2(x\underline{i} + y\underline{j} + z\underline{k})/(x^2 + y^2 + z^2)^2 = -2\underline{r}/r^4$.

17. We want to show that $(d/dt)(f(\underline{\sigma}(t)) = \underline{\nabla}f(\underline{\sigma}(t))\cdot\underline{\sigma}'(t)$. $f_x = (1/2)(2x) \times$
$(x^2 + y^2 + z^2)^{-1/2} = x/\sqrt{x^2 + y^2 + z^2}$. Similarly, we get $\underline{\nabla}f(x,y,z) =$
$(x/\sqrt{x^2 + y^2 + z^2})\underline{i} + (y/\sqrt{x^2 + y^2 + z^2})\underline{j} + (z/\sqrt{x^2 + y^2 + z^2})\underline{k}$, so
$\underline{\nabla}f(\underline{\sigma}(t)) = (\sin t/\sqrt{1 + t^2})\underline{i} + (\cos t/\sqrt{1 + t^2})\underline{j} + (t/\sqrt{1 + t^2})\underline{k}$. Also,
$\underline{\sigma}'(t) = (\cos t, -\sin t, 1)$, and $\underline{\nabla}f(\underline{\sigma}(t))\cdot\underline{\sigma}'(t) = (\sin t \cos t -$
$\cos t \sin t + t)/\sqrt{1 + t^2} = t/\sqrt{1 + t^2}$. On the other hand, $f(\underline{\sigma}(t)) =$
$\sqrt{1 + t^2}$, so the derivative is $(1/2)(1 + t^2)^{-1/2}(2t) = t/\sqrt{1 + t^2}$,
which matches the result above.

21. The directional derivative at $\underline{r}$ in the direction $\underline{d}$ is $\underline{\nabla}f(\underline{r})\cdot\underline{d}$,
where $\underline{d}$ is a unit vector. We have $f_x(1,2) = (2x - 3y^3)|_{(1,2)} =$
$2 - 24 = -22$ and $f_y(1,2) = (2y - 9xy^2)|_{(1,2)} = 4 - 36 = -32$. Then
the directional derivative is $(-22,-32)\cdot(1/2,\sqrt{3}/2) = -11 - 16\sqrt{3}$.

25. Using the method of Exercise 21, we have $f_x(\underline{r}) = (2x - 2y)|_{(1,1,2)} =$
0 , $f_y(\underline{r}) = -2x|_{(1,1,2)} = -2$, and $f_z(\underline{r}) = 6z|_{(1,1,2)} = 12$. Then
the directional derivative is $(0,-2,12)\cdot(1,1,-1)/\sqrt{3} = (0 - 2 - 12)/$
$\sqrt{3} = -14/\sqrt{3}$.

29. The direction of fastest increase occurs in the direction of the gradient.
$f_x(1,1) = 2x|_{(1,1)} = 2$; $f_y(1,1) = 4y|_{(1,1)} = 4$. Hence, $2\underline{i} + 4\underline{j}$ is
the desired direction. $(\underline{i} + 2\underline{j})/\sqrt{5}$ is the unit vector in the desired
direction.

33. (a) Note that $T(1,1,1) = e^{-6}$. Then $T_x(1,1,1) = -2xT|_{(1,1,1)} =$
$-2e^{-6}$; $T_y(1,1,1) = -4yT|_{(1,1,1)} = -4e^{-6}$; and $T_z(1,1,1) =$
$-6zT|_{(1,1,1)} = -6e^{-6}$. The gradient gives the direction of most
rapid increase, so she should proceed in the opposite direction,
i.e., along $(1,2,3)$.

33. (b) Let $\underline{v} = a(1,2,3)$ denote her velocity. Then $e^8 = \|\underline{v}\| = a\sqrt{14}$,

so $a = e^8/\sqrt{14}$. By the chain rule, $dT/dt = T_x(dx/dt) + T_y(dy/dt) +$

$T_z(dz/dt) = (-2e^{-6}e^8/\sqrt{14})(1 + 4 + 9) = -2\sqrt{14}\, e^2$.

(c) We wish to have $dT/dt = -\sqrt{14}e^2$. In terms of the gradient,

$dT/dt = \underline{\nabla}T \cdot \underline{\sigma}'(t)$, where $\underline{\sigma}(t)$ is her trajectory. However,

$\underline{\nabla}T \cdot \underline{\sigma}'(t) = \|\underline{\nabla}T\|\|\underline{\sigma}'(t)\| \cos \theta$. At $(1,1,1)$, this equals

$2e^{-6}(1 + 4 + 9)^{1/2}e^8 \cos \theta = 2\sqrt{14}\, e^2 \cos \theta$, so we need $\cos \theta =$

$1/2$, where θ is the angle between $-\underline{\nabla}T(1,1,1)$ (along $(1,2,3)$ —

see part (a)) and $\sigma'(0)$. Hence $\theta = \pi/6$ and she should fly out-

side the cone with vertex at $(1,1,1)$, axis along $(1,2,3)$, and

sides at an angle of $\pi/6$ from the axis.

37.

$f_x = 2x$ and $f_y = -2y$, so $\underline{\nabla}f(1,0) = 2\underline{i}$,

which is the direction of fastest increase.

41. (a) If $f(x,y,z)$ is constant, then $f_x = f_y = f_z = 0$, so $\underline{\nabla}f = \underline{0}$.

(b) Let $h(x,y,z) = f(x,y,z) + g(x,y,z)$. Then $h_x = f_x + g_x$; $h_y =$

$f_y + g_y$; $h_z = f_z + g_z$, so $\underline{\nabla}(f + g) = \underline{\nabla}h = \underline{\nabla}f + \underline{\nabla}g$.

(c) Let $g(x,y,z) = cf(x,y,z)$. Then $g_x = cf_x$; $g_y = cf_y$; $g_z =$

cf_z , so $\underline{\nabla}(cf) = \underline{\nabla}g = c\underline{\nabla}f$.

(d) Let $h(x,y,z) = f(x,y,z)\, g(x,y,z)$. Then $h_x = f_x g + fg_x$; $h_y =$

$f_y g + fg_y$; $h_z = f_z g + fg_z$. Therefore $\underline{\nabla}(fg) = \underline{\nabla}h = g\underline{\nabla}f + f\underline{\nabla}g$.

(e) Let $h(x,y,z) = f(x,y,z)/g(x,y,z)$, for $g \neq 0$. Then $h_x =$

$(f_x g - fg_x)/g^2$; $h_y = (f_y g - fg_y)/g^2$; $h_z = (f_z g - fg_z)/g^2$.

Therefore, $\underline{\nabla}(f/g) = \underline{\nabla}h = (g\underline{\nabla}f - f\underline{\nabla}g)/g^2$.

45. Let $\underline{\nabla}f(x,y) = (g(x,y),h(x,y))$, $\underline{d}_1$ be the unit vector from the point
    $(1,3)$ to $(2,3)$, i.e., $\underline{d}_1 = (1,0)$; and $\underline{d}_2$ be the unit vector
    from the point $(1,3)$ to $(1,4)$, i.e., $\underline{d}_2 = (0,1)$. We know that
    the directional derivative at $(1,3)$ in the direction $\underline{d}_1$ is
    $\underline{\nabla}f(1,3)\cdot\underline{d}_1 = 2$, i.e., $g(1,3) = 2$. Also $\underline{\nabla}f(1,3)\cdot\underline{d}_2 = -2$, i.e.,
    $h(1,3) = -2$. Hence, $\underline{\nabla}f(1,3) = (g(1,3),h(1,3)) = (2,-2)$. Thus, the
    directional derivative in the direction toward $(3,6)$ is $(2,-2)\cdot\underline{d}_3$,
    where $\underline{d}_3 = [(3,6) - (1,3)]/\|(3,6) - (1,3)\| = (2/\sqrt{13}, 3/\sqrt{13})$. So
    $(2,-2)\cdot\underline{d}_3 = -2/\sqrt{13}$.

49. $\underline{\nabla}f(x,y) = f_x\underline{i} + f_y\underline{j}$, so $\underline{k} \times \underline{\nabla}f = -f_y\underline{i} + f_x\underline{j}$. If this is a gradient,
    then the mixed partials should be equal. Differentiate the $\underline{i}$ compo-
    nent of $\underline{k} \times \underline{\nabla}f$ with respect to y to get $-f_{yy}$. Differentiating the
    $\underline{j}$ component with respect to x gives f_{xx} . Thus, we need $f_{xx} = -f_{yy}$. (The equation $f_{xx} + f_{yy} = 0$ is called Laplace's equation.)
    This is satisfied by any function of the form $ax + bxy + cy + d$,
    where a , b , c , and d are constant. This is also satisfied by
    other special functions such as $f(x,y) = ax^2 + bxy - ay^2 + d$.

SECTION QUIZ

1. Find the direction of fastest decrease for the following:
   (a) $f(x,y) = xy + y^3$ at $(1,2)$
   (b) $f(x,y) = \cos xy + e^x - x$ at $(-1,0)$

2. The directional derivative for $f(x,y) = x^2 + y^2$ in the direction
   $\underline{i} + \underline{j}$ at the point $(1,1)$ is not $(2,2)\cdot(1,1) = 4$.
   (a) Why is the calculation incorrect?
   (b) What is the correct directional derivative?
   (c) Explain what the answer in part (b) tells you geometrically.

3. If $u(x,y) = xy^2 + x^2y + 4$ and $(x,y) = (t + 1, \sin t)$, use the gradient
    to find du/dt .

4. A law-abiding champion downhill skier finds herself on an American
    snow-covered slope whose shape near $(-1,1)$ approximates that of the
    surface $f(x,y) = 25 - x^4 - y^2$. She has determined that her speed
    exceeds the 55 MPH speed limit if the directional derivative at a point
    is less than -2 .

    (a) Does she need to worry about speeding if she skis along the direction
         $-\underline{i} - \underline{j}$? Explain.

    (b) In what direction is her speed greatest?

    (c) In what directions may she ski and still obey the speed limit?

    (d) Can she ski in the same directions as in part (c) and still be
         travelling at less than 55 MPH at $(-2,2)$?

ANSWERS TO PREREQUISITE QUIZ

1. (a) 0

    (b) -6

2. They are orthogonal.

3. $4t^3 \sin t^2 + 2t^5 \cos t^2$

4.

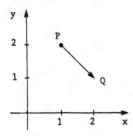

ANSWERS TO SECTION QUIZ

1.   (a) $-2\underline{i} - 13\underline{j}$

(b) $(1 - 1/e)\underline{i}$

2.   (a) $(1,1)$ should be replaced by the unit vector $(1/\sqrt{2}, 1/\sqrt{2})$.

(b) $2\sqrt{2}$

(c)   f rises by $2\sqrt{2}$ units when one goes one unit in the direction

of $\underline{i} + \underline{j}$.

3. $\sin^2 t + 2(t + 1)(\sin t)(1 + \cos t) + (t + 1)^2 \cos t$

4.   (a)   No; directional derivative is $-\sqrt{2}$.

(b) $4\underline{i} - 2\underline{j}$

(c)   Anywhere except the directions from $-\underline{j}$ to $4\underline{i} + 3\underline{j}$,   moving

counterclockwise.

(d)   No.

16.2 Gradients, Level Surfaces, and Implicit Differentiation

PREREQUISITES

1.   Recall how to compute tangent planes to graphs (Section 15.2).

2.   Recall how to compute the gradient of a function (Section 16.1).

3.   Recall how to differentiate implicitly (Section 2.3).

PREREQUISITE QUIZ

1.   Find the equation of the tangent plane to the graph of $z = 2xy + \cos y$
     at the point $(1,0,1)$.

2.   Find the equation of the tangent plane to the graph of $z = 2xy + \cos y$
     at the point $(0, \pi/2, 0)$.

3.   Compute $\underline{\nabla} f$ if $f(x,y) = 2xy + \cos y$.

4.   Compute $\underline{\nabla} f$ if $f(x,y,z) = 3x^2 + yz - z^3$.

5.   Find dy/dx if $x^2 + y^2 = 3 + 2y$.

GOALS

1.   Be able to find a tangent plane by using gradients.

2.   Be able to differentiate a multivariable expression implicitly.

3.   Be able to solve related rate problems involving several variables.

STUDY HINTS

1.   Tangent plane to surface. The tangent plane to a surface at a point is
     that plane containing the tangents to curves which lie in the surface
     and which pass through the point.

2.   Tangent plane normal. The gradient is normal to the tangent plane of
     a level surface. You should definitely know this fact. The equation of
     the tangent plane at $\underline{r}_0$ to the surface $f = $ constant is $\underline{\nabla} f(\underline{r}_0) \cdot (\underline{r} - \underline{r}_0) = $
     0 . Compare this equation with the equation of the plane on p. 672.

3.   **Implicit differentiation.** You should either learn the simple deriva-
     tion just below the box on p. 810 or remember the formula $dy/dx =$
     $-(\partial z/\partial x)/(\partial z/\partial y)$. It looks almost like division of fractions, except
     for the minus sign. Don't forget the minus sign.

4.   **Related rates.** If we know that $F(x,y) =$ constant , then we can dif-
     ferentiate by the chain rule to get $(\partial F/\partial x)(dx/dt) + (\partial F/\partial y)(dy/dt) =$
     0 . The method is analogous to that used in Section 2.4.

SOLUTIONS TO EVERY OTHER ODD EXERCISE

1.

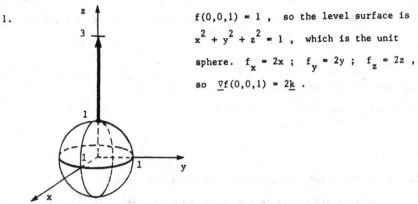

$f(0,0,1) = 1$, so the level surface is
$x^2 + y^2 + z^2 = 1$, which is the unit
sphere. $f_x = 2x$; $f_y = 2y$; $f_z = 2z$,
so $\nabla f(0,0,1) = 2\underline{k}$.

5. $\nabla f(\underline{r}_0)$ is a normal to the surface at $\underline{r}_0$. $f_x = yz$; $f_y = xz$; $f_z =$
     xy , so $\nabla f(1,1,8) = 8\underline{i} + 8\underline{j} + \underline{k}$. $\|\nabla f(1,1,8)\| = \sqrt{129}$, so a unit
     normal is $(1/\sqrt{129})(8\underline{i} + 8\underline{j} + \underline{k})$.

9.   By Example 4, Section 16.1, $\nabla(1/r) = -(\underline{r}/r^3)$. Therefore, we can
     choose $V = Qq/r$ to give $\underline{F} = -\nabla V$.

13. The equation of the tangent plane is $f_x(x_0,y_0,z_0)(x - x_0) + f_y(x_0,y_0,z_0) \times$
     $(y - y_0) + f_z(x_0,y_0,z_0)(z - z_0) = 0$. $\nabla f(x,y,z) = (2x + 3z, 4y, 3x)$,
     which is $(3,8,3)$ at $(1,2,1/3)$. Thus, the tangent plane is $3(x - 1) +$
     $8(y - 2) + 3(z - 1/3) = 0$ or $3x + 8y + 3z = 20$.

17. The equation of the tangent line at (x_0, y_0) to the curve $f(x,y) = c$

is $f_x(x_0, y_0)(x - x_0) + f_y(x_0, y_0)(y - y_0) = 0$. $\underline{\nabla}f(x,y) = (2x, 4y)$.

At $(1,1)$, it becomes $(2,4)$. Thus, the equation for the tangent

line is $2(x - 1) + 4(y - 1) = 0$ or $2x + 4y - 6 = 0 = x + 2y - 3$.

21. The normal to the tangent plane at $\underline{r}_0 = (x_0, y_0, z_0)$ of the level sur-

face $f(x,y,z) = c$ is $\underline{\nabla}f(\underline{r}_0)$, so the normal line is $\underline{r}_0 + t\underline{\nabla}f(\underline{r}_0)$.

$\underline{\nabla}f(x,y,z) = (-2x \exp(-x^2 - y^2 - z^2)$, $-2y \exp(-x^2 - y^2 - z^2)$,

$-2z \exp(-x^2 - y^2 - z^2))$. Since $f_x = f_y = f_z$ at $(1,1,1)$, the

normal line is $(1,1,1) + t(1,1,1)$.

25. Rewrite the given equation as $z = F(x,y) = 0$. Then $dy/dx = -(\partial z/\partial x)/$

$(\partial z/\partial y)$. Let $z = x^2 + 2y^2 - 3$. Then $\partial z/\partial x = 2x$ and $\partial z/\partial y = 4y$;

therefore, $dy/dx = -2x/4y = -x/2y$.

29. Using the method of Exercise 25, we let $z = x^3 - \sin y + y^4 - 4$. Then

$\partial z/\partial x = 3x^2$ and $\partial z/\partial y = -\cos y + 4y^3$; therefore, $dy/dx = 3x^2/$

$(\cos y - 4y^3)$.

33. Let $z = \cos(x + y) - x - 1/2$. Then $\partial z/\partial x = -\sin(x + y) - 1$ and

$\partial z/\partial y = -\sin(x + y)$. At $(0, \pi/3)$, $\partial z/\partial x = -\sqrt{3}/2 - 1$ and $\partial z/\partial y =$

$-\sqrt{3}/2$; therefore, $dy/dx = (\sqrt{3}/2 + 1)/(-\sqrt{3}/2) = -1 - 2\sqrt{3}/3$.

37. $\partial z/\partial y = -5y^4$, which is 0 at $(0,0)$. At that point, the graph of

$x = y^5$ is pointed, so the slope is infinite.

41. Let $F(x,y) = x^4 + y^4 - 1$. Then $\partial F/\partial x = 4x^3$ and $\partial F/\partial y = 4y^3$;

therefore, formula (2) gives $4x^3(dx/dt) + 4y^3(dy/dt) = 0$ or

$x^3(dx/dt) + y^3(dy/dt) = 0$.

45. Let $F(x,y) = \ln(x \cos y) - x = \ln x + \ln \cos y - x$. Then $\partial F/\partial x =$

$1/x - 1$ and $\partial F/\partial y = -\sin y/\cos y = -\tan y$. Therefore, formula (2)

gives $(1/x - 1)(dx/dt) + (-\tan y)(dy/dt) = 0$.

49. The direction normal to the surface is $\underline{\nabla}f(1,1,\sqrt{3})$. We find

$f_x(1,1,\sqrt{3}) = 2x\big|_{(1,1,\sqrt{3})} = 2$; $f_y(1,1,\sqrt{3}) = 2y\big|_{(1,1,\sqrt{3})} = 2$; and

$f_z(1,1,\sqrt{3}) = -2z\big|_{(1,1,\sqrt{3})} = -2\sqrt{3}$, so $\underline{\nabla}f(1,1,\sqrt{3}) = (2,2,-2\sqrt{3})$. Let

$\underline{v} = a(1,1,-\sqrt{3})$ denote the velocity of the particle. Then its speed

$s = \|\underline{v}\| = 10 = a\sqrt{1+1+3} = \sqrt{5}a$, so $a = 2\sqrt{5}$. Hence, $\underline{v} =$

$2\sqrt{5}(1,1,-\sqrt{3})$. It crosses the xy-plane when $z = 0 = \sqrt{3} - 2\sqrt{3}\sqrt{5}t$;

i.e., when $t = 1/2\sqrt{5} = \sqrt{5}/10$. Then $x = 1 + 2\sqrt{5}/2\sqrt{5} = 2 = y$, so it

crosses at $(2,2,0)$, which occurs $\sqrt{5}/10$ seconds later.

SECTION QUIZ

1. Suppose that $2x + 3y + x^2y^2 = 0$ and $y = f(x)$, then $f'(x)$ is <u>not</u>

$\dfrac{-2 + 2xy^2}{3 + 2x^2y}$. What is the common mistake made here which might cost points

on an exam?

2. Find $f'(x)$ if $y = f(x)$ is a function of x :

(a) $y \cos x - \sin(e^y) + x$

(b) $e^{xy} - y \ln xy$

3. Use the gradient to find the tangent plane to the surface $x^2y^2/2 + 3xy =$

$\sqrt{x} - \cos y + z$ at $(1,0,-3/2)$.

4. If $z = F(x,y)$ and x and y are both functions of t , find a

relationship between dx/dt and dy/dt when $xy - z + 2 = 0$.

5. A hungry little bear cub is looking for honey. When he eyes a beehive

which has a paraboloid shape given by $z = 16 - x^2 - y^2$, he quickly

approaches, expecting a delicious breakfast. However, at $(1,2)$, he

is met by an angry swarm of bees. The stinging causes the cub to jump

in a direction normal to the tangent plane of the beehive at $(1,2)$.

(a) What is the direction in which the bear cub springs?

(b) What is the tangent plane of the beehive at $(1,2)$?

ANSWERS TO PREREQUISITE QUIZ

1. $z = 2y + 1$

2. $z = \pi x - y + \pi/2$

3. $2y\underline{i} + (2x - \sin y)\underline{j}$

4. $6x\underline{i} + z\underline{j} + (y - 3z^2)\underline{k}$

5. $x/(1 - y)$

ANSWERS TO SECTION QUIZ

1. The correct answer is $-\left(\dfrac{2 + 2xy^2}{3 + 2x^2y}\right)$. Note parentheses.

2. (a) $(y \sin x - 1)/(\cos x - e^y \cos(e^y))$

    (b) $(y/x - ye^{xy})/(xe^{xy} - \ln xy - 1)$

3. $z = -x/2 + 3y - 1$

4. $dx/dt = -(x/y)dy/dt$

5. (a) $-2\underline{i} - 4\underline{j} - \underline{k}$

    (b) $z = 21 - 2x - 4y$

16.3 Maxima and Minima

PREREQUISITES

1. Recall how to find the local extrema of a function of one variable
    (Section 3.5).

2. Recall how to classify the local extrema of a function of one variable
    by using the first and second derivative tests (Sections 3.2 and 3.3).

PREREQUISITE QUIZ

1. Find the critical points of $f(x) = x^3 + 8x^2 + 5$.

2. If f is continuously differentiable, $f'(x_0) = 0$, and x_0 is a
    local maximum point, what is the sign of $f'(x)$ near x_0 for $x < x_0$
    and $x > x_0$?

3. If x_0 is a local minimum point and $f''(x_0) \neq 0$, what is the sign
    of $f''(x_0)$?

4. Classify the critical points of $f(x) = x^3 - x$.

GOALS

1. Be able to find the extrema of a two-variable function and classify them
    as local minima, local maxima, or saddle points.

STUDY HINTS

1. Definitions. A point is a local minimum if all other points nearby have
    larger values. Notice that this definition is local, referring to nearby
    points. Compare this definition with that in one-variable calculus
    (Section 3.2). A global minimum of f is the smallest value taken by
    f on its entire domain. Similar statements may be made for maxima. An
    extremum (local or global) is either a minimum or maximum.

2. <u>Critical points</u>. For functions of two variables, a critical point is
    any point where <u>both</u> partials vanish (i.e., equal zero). All local ex-
    trema are critical points, but not all critical points are local extre-
    ma. For example, saddle points are critical points, but not local ex-
    trema. Example 6 shows an interesting problem in which the critical
    point is not an extremum.

3. <u>Validity of minimizing squares</u>. In Example 4, we want to solve $\partial d/\partial x =$
    0 and $\partial d/\partial y = 0$. By differentiating d^2 , the chain rule gives
    $2d(\partial d/\partial x)$ and $2d(\partial d/\partial y)$, so we again are solving $\partial d/\partial x = 0$ and
    $\partial d/\partial y = 0$ since $d > 0$.

4. <u>Maximum-minimum test</u>. Consider $Ax^2 + 2Bxy + Cy^2$. If $AC - B^2 < 0$,
    the origin is not a local extremum. If $AC - B^2 > 0$ and $A > 0$, we
    have a local minimum at $(0,0)$. If $AC - B^2 > 0$ and $A < 0$, $(0,0)$
    is a local maximum. This should be memorized. Now, if we let $A = f_{xx}$,
    $B = f_{xy}$, and $C = f_{yy}$, we get the second derivative test on p. 817.

5. <u>Proving the max-min test</u>. In step 1 of the proof of the maximum-minimum
    test for quadratic functions, it is stated that if $AC - B^2 > 0$, $A \neq 0$.
    If $A = 0$, we have $-B^2 > 0$, which is impossible.

6. <u>Finding extrema</u>. If $\partial f/\partial x = 0$ or $\partial f/\partial y = 0$ have more than one solu-
    tion, each combination of (x,y) must be considered. You must be com-
    plete in your analysis. See Example 10.

SOLUTIONS TO EVERY OTHER ODD EXERCISE

1. Notice that the denominator is smallest when $y = 0$; it has no largest
    value, so we expect the minimum and maximum to occur when $y = 0$. This
    should be at a critical point of $x^3 - 3x$, i.e., where $3x^2 - 3 = 0$
    or $x = \pm 1$. From the graph, we see that the minimum is at $(1,0)$ and
    the maximum is at $(-1,0)$.

5. The critical points occur where $\partial f/\partial x$ and $\partial f/\partial y$ both vanish.

    $\partial f/\partial x = (-2x)\exp(-x^2 - 7y^2 + 3)$ and $\partial f/\partial y = (-14y)\exp(-x^2 - 7y^2 + 3)$.

    The partial derivatives vanish at $(0,0)$, so the origin is a critical

    point. It is a local maximum since $-x^2 - 7y^2 \leqslant 0$.

9. Let x and y be the length of the sides of the base. Then the height

    must be $256/xy$. The cost of the box is $C(x,y) = bxy + 2sxz + 2syz$,

    i.e., $A(x,y) = bxy + 512s/y + 512s/x$. The partial derivatives $\partial C/\partial x =$

    $by - (512/x^2)s$ and $\partial C/\partial y = bx - (512/y^2)s$ vanish at $(8 \cdot \sqrt[3]{s/b}, 8 \cdot \sqrt[3]{s/b})$,

    so we have a square based box and the height is $256/xy = 4/\sqrt[3]{s^2/b^2}$

    $= 4b^{2/3}/s^{2/3}$.

13. We have $A = -1$, $B = 3/2$, and $C = 1$, so $AC - B^2 = -13/4 < 0$.

    Therefore, $(0,0)$ is a saddle point.

17. The critical points occur where $\partial f/\partial x = 0 = \partial f/\partial y$. Classify them by

    using the second derivative test. Here, $f_x = 2x + 6$ and $f_y = 2y - 4$,

    so the critical point is at $(-3,2)$. Further, $A = f_{xx} = 2$, $B = f_{xy} =$

    0 , and $C = f_{yy} = 2$. Therefore, $AC - B^2 = 4 > 0$ and $A > 0$, so

    $(-3,2)$ is a local minimum.

21. $f_x = 2x - 6 = 0$ when $x = 3$. $f_y = 2y - 14 = 0$ when $y = 7$. There-

    fore, $(3,7)$ is the critical point. $A = f_{xx} = 2 = f_{yy} = C$ and $B =$

    $f_{xy} = 0$, so $AC - B^2 = 4 > 0$. Since $f_{xx} > 0$, the second deriva-

    tive test tells us that $(3,7)$ is a local minimum.

25. We have $f_x = 6x + 2y - 3$, $f_y = 2x + 4y + 2$; therefore, $B = f_{xy} =$

    2 , $A = f_{xx} = 6$, and $C = f_{yy} = 4$. Solving $f_x = f_y = 0$, we find

    the critical point $(4/5, -9/10)$. Since $A > 0$ and $AC - B^2 > 0$,

    $(4/5, -9/10)$ is a local minimum.

29. $f_x = y \cos xy/(2 + \sin xy) = 0$ implies $y = 0$ or $xy = \pi/2 + n\pi$,

    where n is any integer. Also $f_y = x \cos xy/(2 + \sin xy) = 0$ implies

    $x = 0$ or $xy = \pi/2 + n\pi$. Hence $(0,0)$ and all points satisfying

    $xy = \pi/2 + n\pi$ are critical points.

29.   (continued)

$A = f_{xx} = [-y^2 \sin xy(2 + \sin xy) - (y \cos xy)^2]/(2 + \sin xy)^2 =$

$-y^2(2 \sin xy + \sin^2 xy + \cos^2 xy)/(2 + \sin xy)^2 = -y^2(1 + 2 \sin xy)/$

$(2 + \sin xy)^2$. Similarly, by symmetry, $C = f_{yy} = -x^2(1 + 2 \sin xy)/$

$(2 + \sin xy)^2$. $B = f_{xy} = [(\cos xy - xy \sin xy)(2 + \sin xy) -$

$xy \cos^2 xy]/(2 + \sin xy)^2 = (2 \cos xy - 2xy \sin xy + \cos xy \sin xy -$

$xy \sin^2 xy - xy \cos^2 xy)/(2 + \sin xy)^2 = (2 \cos xy - 2xy \sin xy +$

$\cos xy \sin xy - xy)/(2 + \sin xy)^2$. If $(x,y) = (0,0)$, we get $A =$

$0 = C$ and $B = 1/2$, so $AC - B^2 = -1/4 < 0$. Thus $(0,0)$ is a

saddle point.

33.

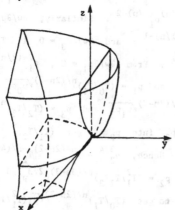

$f_x = 2x = 0$ at $(0,0)$. $f_y = 3y^2 = 0$ at $(0,0)$. $f_{xx} = 2$; $f_{yy} = 6y$; and $f_{xy} = 0$. Hence, $f_{xx}f_{yy} - f_{xy}^2 = 0$ at $(0,0)$, meaning the second derivative test fails. Plot the graph. When $y = 0$, the level curve is the parabola $z = x^2$. When $x = 0$, the level curve is the cubic $z = y^3$, which has an inflection point at $(0,0)$. When $z = 0$, the level curve is described by $y = -x^{2/3}$. Hence, $(0,0)$ must be a saddle point.

37.   (a) Minimize w by setting $\partial w/\partial p_i = 0$. Since $w = c_1 y + c_2$,

$\partial w/\partial p_i = c_1(\partial y/\partial p_i) = c_1(\partial/\partial p_i)(\Sigma T_i(p_i/p_{i+1})^{(n-1)/n})$. Since only

two terms involve p_i , we have $c_1(\partial/\partial p_i)(T_{i-1}(p_{i-1}/p_i)^{(n-1)/n} +$

$T_i(p_i/p_{i+1})^{(n-1)/n}) = c_1(\partial/\partial p_i)(T_{i-1}p_{i-1}^{1-(1/n)} p_i^{(1/n)-1} + T_i^{1-(1/n)}p_i p_{i+1}^{(1/n)-1}) =$

$c_1[T_{i-1}p_{i-1}^{1-(1/n)}(1/n - 1)p_i^{(1/n)-2} + T_i p_{i+1}^{(1/n)-1}(1 - 1/n)p_i^{-1/n}] =$

$c_1(1 - 1/n)[T_i p_{i+1}^{(1/n)-1}p_i^{1/n} - T_{i-1}p_{i-1}^{1-(1/n)} p_i^{(1/n)-2}]$. Hence,

37. (a) (continued)

$\partial w/\partial p_i = 0$ if $T_i p_{i+1}^{(1/n)-1} p_i^{-1/n} = T_{i-1} p_{i-1}^{1-(1/n)} p_i^{(1/n)-2}$. Solve

for $T_{i-1}/T_i = p_{i+1}^{(1/n)-1} p_i^{-1/n} p_{i-1}^{(1/n)-1} p_i^{2-(1/n)} = p_{i+1}^{(1/n)-1} p_i^{2-(2/n)} \times$

$p_{i-1}^{(1/n)-1} = (p_i^2/(p_{i-1}p_{i+1}))^{1-(1/n)}$, which is the necessary

condition to minimize w .

(b) $y = T_0(p_0/p_1)^{1-(1/n)} + T_1(p_1/p_2)^{1-(1/n)} + K$, where $K = T_2 =$

$(p_2/p_3)^{1-(1/n)} + T_3(p_3/p_4)^{1-(1/n)}$. Then $y = T_0 p_0^{1-(1/n)} p_1^{(1/n)-1} +$

$T_1 p_1^{1-(1/n)} p_2^{(1/n)-1} + K$. Consequently, $\partial w/\partial p_1 = c_1 [T_0 p_0^{1-(1/n)} \times$

$(1/n - 1)p_1^{(1/n)-2} + T_1 p_2^{(1/n)-1}(1 - 1/n)p_1^{-1/n}] = c_1(1 - 1/n) \times$

$[T_1 p_2^{(1/n)-1} p_1^{-1/n} - T_0 p_0^{1-(1/n)} p_1^{(1/n)-2}]$. Hence, $\partial w/\partial p_1 = 0$ if

$T_1 p_2^{(1/n)-1} p_1^{-1/n} = T_0 p_0^{1-(1/n)} p_1^{(1/n)-2}$. Similarly, $\partial w/\partial p_2 = 0$ if

$T_1 = T_2 p_3^{(1/n)-1} p_2^{2-(2/n)} p_1^{(1/n)-1}$; and $\partial w/\partial p_3 = 0$ if $T_2 =$

$T_3 p_4^{(1/n)-1} p_3^{2-(2/n)} p_2^{(1/n)-1}$. From $\partial w/\partial p_1 = 0$, $p_1^{2-(2/n)} =$

$(T_0/T_1)p_2^{1-(1/n)} p_0^{1-(1/n)}$, so $p_1 = (T_0/T_1)^{n/(2n-2)}\sqrt{p_0 p_2}$.

Similarly, $p_2 = (T_1/T_2)^{n/(2n-2)}\sqrt{p_1 p_3}$ and $p_3 = (T_2/T_3)^{n/(2n-2)} \times$

$\sqrt{p_2 p_4}$. Substitute for p_3 into p_2 to get $(T_1/T_2)^{n/(2n-2)} \times$

$\sqrt{p_1}(T_2/T_3)^{n/(4n-4)}\sqrt{p_2 p_4}$. Hence, $p_2^{3/4} = T_1^{n/(2n-2)} T_2^{-n/(4n-4)} \times$

$T_3^{-n/(4n-4)}\sqrt{p_1} \sqrt[4]{p_4}$, so $p_2 = (T_1^2/T_2 T_3)^{n/(3n-3)} p_1^{2/3} p_4^{1/3}$.

Substitute this into p_1 to get $(T_0/T_1)^{n/(2n-2)}\sqrt{p_0}(T_1^2/T_2 T_3)^{n/(6n-6)} \times$

$p_1^{1/3} p_4^{1/6}$. Hence, $p_1^{2/3} = \sqrt{p_0} p_4^{1/6} T_0^{n/(2n-2)} T_1^{-n/(6n-6)}(T_2 T_3)^{-n/(6n-6)}$.

Thus, $p_1 = p_0^{3/4} p_4^{1/4} T_0^{3n/(4n-4)} T_1^{-n/(4n-4)}(T_2 T_3)^{-n/(4n-4)} = [p_0^3 p_4(T_0^3/$

$T_1 T_2 T_3)^{n/(n-1)}]^{1/4}$. Substitute this into p_2 to get

$(T_1^2/T_2 T_3)^{n/(3n-3)} p_4^{1/3} \cdot [p_0^3 p_4(T_0^3/T_1 T_2 T_3)^{n/(n-1)}]^{1/6} =$

$p_0^{1/2} p_4^{1/2} T_0^{n/(2n-2)} T_1^{n/(2n-2)} T_2^{-n/(2n-2)} T_3^{-n/(2n-2)} = [p_0 p_4(T_0 T_1/$

$T_2 T_3)^{n/(n-1)}]^{1/2}$. Substitute this into p_3 to get $p_3 = T_2^{n/(2n-2)} \times$

$T_3^{-n/(2n-2)} p_4^{1/2} \cdot [p_0 p_4(T_0 T_1/T_2 T_3)^{n/(n-1)}]^{1/4} = p_0^{1/4} p_4^{3/4}(T_0 T_1)^{n/(4n-4)}$

$T_2^{n/(4n-4)} T_3^{-3n/(4n-4)} = [p_0 p_4^3(T_0 T_1 T_2/T_3^3)^{n/(n-1)}]^{1/4}$.

41. (a) $f_x = 2x + 3y$ and $f_y = -2y + 3x$. At $(0,0)$, $f_x = f_y = 0$. $f_{xx} = 2$; $f_{yy} = -2$; and $f_{xy} = 3$. Hence $f_{xx}f_{yy} - f_{xy}^2 = -4 - 9 = -13 < 0$. The second derivative test tells us that $(0,0)$ is a saddle point.

(b) $f_x = 2x + Cy$ and $f_y = 2y + Cx$. At $(0,0)$, $f_x = f_y = 0$. $f_{xx} = 2$; $f_{yy} = 2$; and $f_{xy} = C$, so $f_{xx}f_{yy} - f_{xy}^2 = 4 - C^2$. Hence if $-2 < C < 2$, $(0,0)$ is a strict minimum; but if $C < -2$ or $C > 2$, then $(0,0)$ is a saddle point. If $C = \pm 2$, then $f(x) = (x \pm y)^2$, so $(0,0)$ is a minimum. The behavior changes qualitatively at $C = \pm 2$.

45. $s(m,b) = \Sigma(y_i - mx_i - b)^2$. Therefore $\partial s/\partial b = \Sigma(-2)(y_i - mx_i - b)$. If $\partial s/\partial b = 0$, then $\Sigma(y_i - mx_i - b) = 0$. Using the rules for summation (See Section 4.1), $\Sigma y_i - m\Sigma x_i - nb = 0$, so $m\Sigma x_i + nb = \Sigma y_i$. Also, $\partial s/\partial m = \Sigma(-2x_i)(y_i - mx_i - b)$. If $\partial s/\partial m = 0$, then $\Sigma x_i(y_i - mx_i - b) = 0$, so $\Sigma x_i y_i = m\Sigma x_i^2 + b\Sigma x_i$.

49. (a) The test gives $g(x,y) = Ax^2 + 2Bxy + Cy^2 = A[x^2 + 2Bxy/A + Cy^2/A] = A[x^2 + 2Bxy/A + (By/A)^2 - (By/A)^2 + Cy^2/A] = A[(x + By/A)^2 - y^2(B^2 - CA)/A^2]$. Let $e = \sqrt{B^2 - CA}/A$. Then $g(x,y) = A[(x + By/A)^2 - y^2e^2] = A[(x + (By/A) - ey)(x + (By/A) + ey)]$.

(b) $g(x,y) = 0$ if $x + By/A - ey = 0$ or if $x + By/A + ey = 0$. Solve these for y as follows: $x + y(B/A - e) = 0$ implies $y = x/(e - B/A) = Ax/(Ae - B)$; and $x + y(B/A + e) = 0$ implies $y = -x/(B/A + e) = -Ax/(B + Ae)$. These are lines that intersect at the origin.

(c) $g(x,y)$ is positive if $[x + By/A - ey > 0$ and $x + By/A + ey > 0]$ or $[x + By/A - ey < 0$ and $x + By/A + ey < 0]$. This reduces to $[y > Ax/(Ae - B)$ and $y > -Ax/(B + Ae)]$ or $[y < Ax/(Ae - B)$ and $y < -Ax/(B + Ae)]$. These are the regions above both lines and below both lines, respectively.

49. (d) If $A = B = 0$, then $AC - B^2 = 0$ and the test does not apply.

Rewrite $g(x,y)$ as $y(2Bx + Cy)$. Note that $y = 0$ and $2Bx + Cy = 0$ are two lines intersecting at the origin. Thus, $g(x)$ is positive in the region above $2Bx + Cy = 0$ and the negative x-axis. Also, $g(x)$ is positive in the region below $2Bx + Cy = 0$ and the positive x-axis.

53. As in Example 10, we minimize the square of the distance: $f(x,y) = x^2 + y^2 + (4x^2 + y^2 - a)^2$. The partials are $f_x(x,y) = 2x + 2(4x^2 + y^2 - a)8x = 2x(32x^2 + 8y^2 - 8a + 1)$ and $f_y(x,y) = 2y + 2(4x^2 + y^2 - a)2y = 2y(8x^2 + 2y^2 - 2a + 1)$. Thus, we need $x = 0$ or $32x^2 + 8y^2 - 8a + 1 = 0$. Also, we need $y = 0$ or $8x^2 + 2y^2 - 2a + 1 = 0$. Again, there are four possibilities. $(0,0)$ is one critical point.

For $x = 0$ and $8x^2 + 2y^2 - 2a + 1 = 0$, we have $2y^2 - 2a + 1 = 0$ or $y = \pm\sqrt{(2a - 1)/2}$.

For $y = 0$ and $32x^2 + 8y^2 - 8a + 1 = 0$, we have $32x^2 - 8a + 1 = 0$ or $x = \pm\sqrt{(8a - 1)/32}$.

For $32x^2 + 8y^2 - 8a + 1 = 0$ and $8x^2 + 2y^2 - 2a + 1 = 0$, subtracting 4 times the second equation from the first gives $-3 = 0$, which is impossible.

The second derivatives are $f_{xx}(x,y) = 2(32x^2 + 8y^2 - 8a + 1) + 2x(64x) = 192x^2 + 16y^2 - 16a + 2$, and $f_{yy}(x,y) = 2(8x^2 + 2y^2 - 2a + 1) + 2y(4y) = 16x^2 + 12y^2 - 4a + 2$, and $f_{xy}(x,y) = 32xy$.

At $(x,y) = (0,0)$, we have $f_{xx} = -16a + 2$, $f_{yy} = -4a + 2$, and $f_{xy} = 0$. Thus, $f_{xx}f_{yy} - f_{xy}^2 = 64a^2 - 40a + 4$. At $(x,y) = (0,\pm\sqrt{(2a - 1)/2})$, we have $f_{xx} = -6$, $f_{yy} = 8a - 4$, and $f_{xy} = 0$.

53. (continued)

Thus, $f_{xx}f_{yy} - f_{xy}^2 = -48a + 24$. At $(x,y) = (\pm\sqrt{(8a - 1)/32},0)$, we have $f_{xx} = 32a - 4$, $f_{yy} = 3/2$, and $f_{xy} = 0$. Thus, $f_{xx}f_{yy} - f_{xy}^2 = 48a - 6 = 6(8a - 1)$.

Therefore, if $a < 1/8$, $(0,0)$ is the only critical point, $f_{xx} > 0$, and $f_{xx}f_{yy} - f_{xy}^2 > 0$, so it is a local minimum. If $a = 1/8$, the critical point is $(0,0)$ and it is a local minimum. If $1/8 < a < 1/2$, then the critical points are $(0,0)$ and $(\pm\sqrt{(8a - 1)/32},0)$. In this case, the second derivative test tells us that $(\pm\sqrt{(8a - 1)/32},0)$ gives the minimum. If $a = 1/2$, the second derivative test fails for $(0,0)$ and tells us that $(\pm\sqrt{3/32},0)$ is a local minimum. Substituting back into $f(x,y)$ tells us it is the global minimum. If $a > 1/2$, all three points are critical points and the second derivative test still tells us that $(\pm\sqrt{(8a - 1)/32},0)$ is the minimum. Thus, $(0,0,0)$ is the closest point if $a \leqslant 1/8$ and $(\pm\sqrt{(8a - 1)/32},0),(8a - 1)/8)$ are the closest points if $a \geqslant 1/8$.

SECTION QUIZ

1. Consider a two-variable function, $f(x,y)$. Explain how $f_{xx}f_{yy} - [f_{xy}]^2$ is used to classify critical points of f .

2. Find the critical points of $f(x,y)$ and classify them:

   (a) $f(x,y) = x^2 - 5xy + 8y^2$

   (b) $f(x,y) = e^{xy} + 3x - 3y$ [Hint: A calculator and a graph may be useful.]

   (c) $f(x,y) = xy - x^2 + y^2$

   (d) $f(x,y) = 1 - 2y^2$

3.   Find and classify all critical points of $z = (y^3/3 - y)(x^2/2 - 2)$.

4.   A sadistic dentist likes to explain to his patients how they can expect

to feel pain. He explains that pain is a function of   c , the dental

bill which the patient pays without the help of insurance; t , the time

spent by the patient trying to understand the dentist's pain formula;

and n , the number of teeth to be yanked out. This dentist points out

that his patient's pain, P , is given by $P = c^2 t + 2^n$. Explain why

P has no critical points. (Assume that P is a continuous function of

n because this dentist sometimes mistakenly pulls out only parts of a

tooth.)

ANSWERS TO PREREQUISITE QUIZ

1. $-16/3, 0$

2. $f'(x_0) < 0$ for $x < x_0$ and x near x_0 ; $f'(x_0) > 0$ for $x > x_0$

and x near x_0 .

3.   Positive

4. $-\sqrt{1/3}$ = local maximum; $+\sqrt{1/3}$ = local minimum.

ANSWERS TO SECTION QUIZ

1.   Let (x_0, y_0) be a critical point and let $A = f_{xx}(x_0, y_0)$, $B =$
$f_{xy}(x_0, y_0)$, and $C = f_{yy}(x_0, y_0)$. If $AC - B^2 > 0$ and $A > 0$,
then (x_0, y_0) is a local minimum. If $AC - B^2 > 0$ and $A < 0$, then
(x_0, y_0) is a local maximum. If $AC - B^2 < 0$, then (x_0, y_0) is
a saddle point. If $AC - B^2 = 0$, the test is inconclusive.

2.   (a) $(0,0)$ is a local minimum; $f(x,y) = (x + 5y/2)^2 + 7y^2/4$.

(b) There are no critical points.

(c) $(0,0)$ is a saddle point.

(d) $(x,0)$ is a local maximum.

3. $(0,1)$ is a local maximum; $(0,-1)$ is a local minimum; $(\pm 2,0)$,
      $(\pm 2,\sqrt{3})$ and $(\pm 2,-\sqrt{3})$ are all saddle points.

4. $\partial P/\partial n = 2^n(\ln 2) \neq 0$ for any n. A critical point must satisfy
      $\partial P/\partial c = \partial P/\partial t = \partial P/\partial n = 0$.

16.4 Constrained Extrema and Lagrange Multipliers

PREREQUISITES

1.   Recall the extreme value theorem (Section 3.5)

2.   Recall how to find the local extrema of a function of several
     variables (Section 16.3).

3.   Recall how to solve minimum-maximum word problems (Section 3.5).

PREREQUISITE QUIZ

1.   Find the minimum and maximum of $f(x) = x^2$ on $[-1,3]$.

2.   If 12 meters of rope are available to rope off a region, what is
     the maximum area of the region?

3.   State the extreme value theorem.

4.   If $f(x,y)$ is a function of two variables, what must be the value
     of $\partial f/\partial x$ and $\partial f/\partial y$ at a local extreme point?

GOALS

1.   Be able to use the method of Lagrange multipliers to find the minimum
     and maximum for functions with constraints.

2.   Be able to find the minimum and maximum of a two-variable function on
     a region with a boundary.

STUDY HINTS

1.   Constraints. In Chapter 3, we limited our analysis to parts of the
     real line by specifying endpoints. Analogously, a boundary curve
     limits the region of our analysis in the plane.

2.   Locating boundary extrema. Two methods are used. The first is to parametrize the given boundary and then differentiate to determine the local minimum and local maximum points. See Step 2 of Example 1. The second method is that of Lagrange multipliers.

3.   Lagrange multipliers. Rather than remembering the equations (1), there is another way to get the same result. We transform the constraint $g(x,y) = c$ into $g(x,y) - c = 0$. Being sure that the constraint equals zero, we look for critical points of $f(x,y) - \lambda[g(x,y) - c]$, i.e., the original function minus λ times the constraint.

4.   Extrema on a region. The two methods discussed above only locate extrema on a boundary. Don't forget to analyze the critical points inside the region.

SOLUTIONS TO EVERY OTHER ODD EXERCISE

1.   Use the method of Example 1. $\partial f/\partial x = 4x$ and $\partial f/\partial y = 6y$, so $(0,0)$ is the only critical point. On the boundary, we have $h(t) = 2\cos^2 t + 3\sin^2 t$. Thus, $h'(t) = -4\cos t \sin t + 6\sin t \cos t = 2\sin t \cos t = \sin 2t$. The possible extrema points may occur at $t = \pi/4, 3\pi/4, 5\pi/4, 7\pi/4$. At each of these points, $(\cos t, \sin t) = (\pm\sqrt{2}/2, \pm\sqrt{2}/2)$. Thus, the minimum value of 0 occurs at $(0,0)$ and the maximum occurs at four boundary points; the maximum value is $5/2$.

5.   Use the method of Lagrange multipliers. We want to analyze $k(x,y,\lambda) = 3x + 2y + \lambda(2x^2 + 3y^2 - 3)$. $k_x = 3 + 4x\lambda$, $k_y = 2 + 6y\lambda$, and $k_\lambda = 2x^2 + 3y^2 - 3$. Setting the partials equal to 0, we get $x = -3/4\lambda$, $y = -1/3\lambda$; therefore, $k_\lambda = 9/8\lambda^2 + 1/3\lambda^2 - 3 = 0 = 35/24\lambda^2 - 3$. Thus, $\lambda = \pm\sqrt{35/72}$, $x = \mp 9/\sqrt{70}$, and $y = \mp\sqrt{8/35}$. Since $f_x = 3$ and $f_y = 2$, there are no critical points. $f(9/\sqrt{70}, \sqrt{8/35}) = \sqrt{35/2}$ and

5. (continued)

$f(-9/\sqrt{70}, -\sqrt{8/35}) = -\sqrt{35/2}$, which are the extreme values.

9. We want to analyze $k(x,y,\lambda) = xy + \lambda(x + y - 1)$. $k_x = y + \lambda$; $k_y = x + \lambda$; $k_\lambda = x + y - 1$. $k_x = k_y = k_\lambda = 0$ implies $y = -\lambda = x$, so $-2\lambda - 1 = 0$; i.e., $\lambda = -1/2$. Then $x = y = 1/2$. $f(1/2, 1/2) = 1/4$. Hence, $(1/2, 1/2)$ is the extreme point. Consider $f(0,1) = 0$. Since $(1/2, 1/2)$ is an extreme point and $f(0,1) < f(1/2, 1/2)$, the point $(1/2, 1/2)$ must be the maximum point, with value $1/4$.

13. Use Lagrange multipliers. Let $f(x,y,z,\lambda) = 8xyz^2 - 200{,}000(x + y + z) + \lambda(x + y + z - 100{,}000)$. Then $f_x = 8yz^2 - 200{,}000 + \lambda$; $f_y = 8xz^2 - 200{,}000 + \lambda$; $f_z = 16xyz - 200{,}000 + \lambda$; $f_\lambda = x + y + z - 100{,}000$. Since $f_x = 0$, $\lambda = 200{,}000 - 8yz^2$. Since $f_y = 0$, $\lambda = 200{,}000 - 8xz^2$. Therefore $x = y$. Since $f_z = 0$, $\lambda = 200{,}000 - 16xyz = 200{,}000 - 16x^2z$. Therefore, $8xz^2 = 16x^2z$, so $z = 2x$ (since $x \neq 0$, $z \neq 0$). Substitute for y and z in $f_\lambda = 0$, giving $4x = 100{,}000$. Hence $x = y = 25{,}000$ and $z = 50{,}000$.

17. Let $k(D_1, D_2, \lambda) = \ell_1(a + bD_1) + \ell_2(a + bD_2) + \lambda(c\ell_1 Q_1^2/D_1^5 + c\ell_2 Q_2^2/D_2^5 - h)$. Then $\partial k/\partial D_1 = \ell_1 b - 5\lambda c\ell_1 Q_1^2/D_1^b = \ell_1(b - 5\lambda cQ_1^2/D_1^6)$; $\partial k/\partial D_2 = \ell_2(b - 5\lambda cQ_2^2/D_2^6)$; and $\partial k/\partial\lambda = c\ell_1 Q_1^2/D_1^5 + c\ell_2 Q_2^2/D_2^5 - h$. If $\partial k/\partial D_1 = 0$, then $\lambda = bD_1^6/5cQ_1^2$. If $\partial k/\partial D_2 = 0$, then $\lambda = bD_2^6/5cQ_2^2$. Equate these to get $D_2/D_1 = (Q_2/Q_1)^{1/3}$.

21. (a) $n_1 i_1^2/q_1 = n_1 i_1^2/\alpha xh$. Since $n_1 i_1$ is constant, so is $n_1^2 i_1^2/\alpha h$. Call this constant K . Similarly, $n_2 i_2^2/q^2 = K/y$. Thus, C simplifies to $\rho\pi K[(D_1 + x)/x + (D_2 - y)/y] = \rho\pi K(D_1/x + D_2/y)$.

(b) To minimize C , we look at $f(x,y,\lambda) = \rho\pi K(D_1/x + D_2/y) + \lambda(x + y - (1/2)(D_2 - D_1))$. $f_x = -\rho\pi KD_1/x^2 + \lambda$, $f_y = -\rho\pi KD_2/y^2 + \lambda$ and $f_\lambda = x + y - (1/2)(D_2 - D_1)$. $f_x = f_y = 0$ implies $D_1/x^2 =$

21. (b) (continued)

D_2/y^2 , so $y = \sqrt{D_2/D_1}\,x$. $f_\lambda = 0$ implies $(1 + \sqrt{D_2/D_1})x -$
$(1/2)(D_2 - D_1) = 0$, so $x = (D_2 - D_1)/2(1 + \sqrt{D_2/D_1})$ and $y =$
$(D_2 - D_1)\sqrt{D_2/D_1}/2(1 + \sqrt{D_2/D_1})$.

SECTION QUIZ

1. The function $f(x,y) = 1/xy + x + y$ has local extrema, yet it has no
    extreme values in the disk $x^2 + y^2 \leq 2$.

    (a) Why does this function with constraints not have extreme values?

    (b) Find the local extrema points.

2. The function xy^2 has no critical points in the first quadrant, but
    extreme values exist in the rectangular region $1 \leq x \leq 3$ and $3 \leq y \leq 4$.
    Find the extreme values.

3. Find the minimum and maximum of $x^2 y^2$ in the region bounded by $y = x^2$,
    $y = -\sqrt{1 - x^2}$, and $x = \pm 1$.

4. A certain little gopher's weight gain depends on how much vegetables
    the Green family plants. Due to the shape of the terrain, the vegetables
    at point (x,y) provide $36 - x^2$ weight units for the gopher. The
    sprinkler system provides easy passage parallel to the x-axis only, so
    the weight gain is given by $W = 36 - x^2 - y$. All of the Greens'
    vegetables are planted in the disk $x^2 + y^2 \leq 4$.

    (a) Where does the gopher minimize his weight gain, W , if he eats
         the Greens' vegetables?

    (b) At what spot should the gopher eat to become as chubby as possible?

ANSWERS TO PREREQUISITE QUIZ

1.   Minimum = 0 ; maximum = 9 .

2.   9 m^2

3.   A minimum and maximum must exist if (i) the graph is continuous and
     (ii) the domain is closed.

4. $\partial f/\partial x = \partial f/\partial y = 0$

ANSWERS TO SECTION QUIZ

1.   (a) It is not continuous at (0,0) .

     (b) (1,1,3)

2.   Minimum = 3 ; maximum = 48 .

3.   Minimum = 0 on x-axis (-1 $\leqslant$ x $\leqslant$ 1) and on y-axis (-1 $\leqslant$ y $\leqslant$ 0) ;
     maximum = 1 at (±1,1) .

4.   (a) $(-\sqrt{15}/2, 1/2)$

     (b) (0,-2)

16.R <u>Review Exercises for Chapter 16</u>

SOLUTIONS TO EVERY OTHER ODD EXERCISE

1. The gradient is $(\partial f/\partial x)\underline{i} + (\partial f/\partial y)\underline{j}$. $\underline{\nabla}f(x,y) = [y \exp(xy) -$
    $y \sin(xy)]\underline{i} + [x \exp(xy) - x \sin (xy)]\underline{j}$.

5. (a) Since $\underline{d}$ is a unit vector, the directional derivative is $\underline{\nabla}f \cdot \underline{d}$.
    $\underline{\nabla}f = 3x^2 \cos(x^3 - 2y^3)\underline{i} - 6y^2\cos(x^3 - 2y^3)\underline{j}$. At $(1,-1)$, $\underline{\nabla}f =$
    $3 \cos (3)\underline{i} - 6 \cos (3)\underline{j}$; therefore, the directional derivative
    is $(3/\sqrt{2})\cos(3) + (6/\sqrt{2})\cos(3) = (9/\sqrt{2})\cos(3)$.

    (b) The direction of fastest increase is $\underline{\nabla}f = 3 \cos (3)\underline{i} -$
    $6 \cos (3)\underline{j}$ or $(\underline{i} - 2\underline{j})/\sqrt{5}$.

9. The normal to the tangent plane is $\underline{\nabla}f(x_0,y_0,z_0)$. Rewrite the surface
    as $x^3 + 2y^2 - z = 0$, so $\underline{\nabla}f = (3x^2,4y,-1)$. Thus, the tangent plane
    is $3(x - 1) + 4(y - 1) - (z - 3) = 0$ or $3x + 4y - z = 4$.

13. The relationship we use is $F_x(x,y)(dx/dt) + F_y(x,y)(dy/dt) = 0$.
    $F_x = 2x + y$ and $F_y = x + 2y$, so $(2x + y)(dx/dt) + (x + 2y)(dy/dt) = 0$.

17. Use the formula $dy/dx = -(\partial F/\partial x)/(\partial F/\partial y)$. $\partial F/\partial x = 1$ and $\partial F/\partial y =$
    $-\sin y$, so at $(1,\pi/2)$, $dy/dx = -1/(-1) = 1$.

21. The critical points occur where $\partial f/\partial x = \partial f/\partial y = 0$. Use the second
    derivative test to classify the critical points. $\partial f/\partial x = 2x - 6y$ and
    $\partial f/\partial y = -6x - 2y$. The only critical point is $(0,0)$. $\partial^2 f/\partial x^2 = 2$,
    $\partial^2 f/\partial y^2 = -2$, and $\partial^2 f/\partial x \partial y = -6$. Thus, $AC - B^2 = -32 < 0$, so
    $(0,0)$ is a saddle point.

25. $z_x = (12x^3 - 12x^2 - 24x)/12(1 + 4y^2) = x(x^2 - x - 2)/(1 + 4y^2) =$
    $x(x - 2)(x + 1)/(1 + 4y^2) = 0$ if $x = -1$, 0 , or 2 . Also, $z_y =$
    $-2y(3x^4 - 4x^3 - 12x^2 + 18)/3(1 + 4y^2)^2 = 0$ if $y = 0$. Therefore
    critical points occur at $(-1,0)$, $(0,0)$, and $(2,0)$. Further,

25. (continued)

$z_{xx} = (3x^2 - 2x - 2)/(1 + 4y^2)$; $z_{yy} = (-2/3)(3x^4 - 4x^3 - 12x^2 + 18) \times$

$[(1 + 4y^2)^{-2} - 2(1 + 4y^2)^{-3}(8y^2)] = (-2/3)(3x^4 - 4x^3 - 12x^2 + 18)(1 +$

$4y^2 - 16y^2)/(1 + 4y^2)^3 = (-2/3)(3x^4 - 4x^3 - 12x^2 + 18)(1 - 12y^2)/$

$1 + 4y^2)^3$; and $z_{xy} = -x(x - 2)(x + 1)(8y)/(1 + 4y^2)^2$. At $(-1,0)$,

$z_{xx} = (3 + 2 - 2)/1 = 3$; $z_{yy} = (-2/3)(3 + 4 - 12 + 18) = (-2/3)(13) =$

$-26/3$; and $z_{xy} = 0$. Hence $z_{xx}z_{yy} - z_{xy}^2 = -26 < 0$, so $(-1,0)$

is a saddle point. At $(0,0)$, $z_{xx} = -2$; $z_{yy} = -12$; and $z_{xy} = 0$,

so $z_{xx}z_{yy} - z_{xy}^2 = 24 > 0$, so $(0,0)$ is a local maximum point. Finally,

at $(2,0)$, $z_{xx} = 12 - 8 - 2 = 2$; $z_{yy} = (-2/3)(48 - 32 - 48 + 18) =$

$(-2/3)(-14) = 28/3$; and $z_{xy} = 0$. Hence $z_{xx}z_{yy} - z_{xy}^2 = 56/3 > 0$,

so $(2,0)$ is a local minimum point.

29. (a) $\|\underline{G}\| = ((\partial P/\partial x)^2 + (\partial P/\partial y)^2)^{1/2}$.

(b) $\underline{G}$ creates a force on the air mass, which produces a proportionate

acceleration of the air mass in the same direction as $\underline{G}$,

according to Newton's second law of motion.

(c)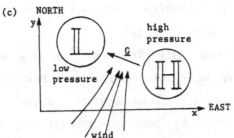

This rotation of the earth turns the wind direction away from $\underline{G}$.

(d)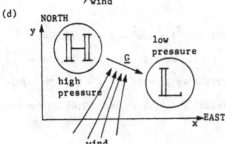

If, in the Southern Hemisphere, you stand with your back to the wind, the high pressure is on your left and the low pressure on your right.

33. It is easier to parametrize the circle. Then $f(x,y)$ becomes $h(t) =$ $\cos^2 t - 2 \sin t \cos t + 2 \sin^2 t = 1 - 2 \sin t \cos t + \sin^2 t$. $h'(t) =$ $-2 \cos^2 t + 2 \sin^2 t + 2 \sin t \cos t = -2 \cos 2t + \sin 2t$. This is 0 if $2 \cos 2t = \sin 2t$, i.e., $2 = \tan 2t$. Thus, $t = 0.554$, 2.124 , 3.695 , and 5.266 . $h(0.554) = h(2.124) \approx 0.382$ and $h(2.124) =$ $h(5.266) \approx 2.618$. These are the two extreme values.

37. Use Lagrange multipliers. Let $f(x,y,\lambda) = x + 2y \sec \theta + \lambda(xy + y^2 \tan \theta - A)$. Then $f_x = 1 + \lambda y$; $f_y = 2 \sec \theta + \lambda(x + 2y \tan \theta)$; and $f_\lambda = xy + y^2 \tan \theta - A$. From $f_x = 0$, we get $\lambda = -1/y$; whereas from $f_y = 0$, we get $\lambda = -2 \sec \theta/(x + 2y \tan \theta)$. Hence $2y = (x + 2y \tan \theta)/\sec \theta$, so $2y(1 - \sin \theta) = x \cos \theta$. Thus, $x = 2y(\sec \theta - \tan \theta)$. Substitute this into $f_\lambda = 0$ to get $2y^2(\sec \theta - \tan \theta) + y^2 \tan \theta = A$. Then $y^2(2 \sec \theta - \tan \theta) = A$, so $y^2 = A/(2 \sec \theta - \tan \theta) = A \cos \theta/(2 - \sin \theta)$.

41. Let P_1 and P_2 denote the desired tangent planes.

(a) At $(1,1,2)$, $f_x = 2x = 2$; $f_y = 2y = 2$; $f_z = 2z = 4$. Hence P_1 is $2(x - 1) + 2(y - 1) + 4(z - 2) = 0$; i.e., $x + y + 2z = 6$. One normal vector is $(1,1,2)$. $g_x = 4x = 4$; $g_y = 6y = 6$; $g_z = 2z = 4$; so P_2 is $4(x - 1) + 6(y - 1) + 4(z - 2) = 0$. Rewrite it as $2x + 3y + 2z = 9$. Then one normal vector is $(2,3,2)$.

(b) Let θ denote the angle between P_1 and P_2 . Then $(1,1,2) \cdot (2,3,2) = \| (1,1,2) \| \| (2,3,2) \| \cos \theta$. Solve for $\theta = \cos^{-1}((2 + 3 + 4)/\sqrt{(1 + 1 + 4)(4 + 9 + 4)}) = \cos^{-1}(9/\sqrt{6 \cdot 17}) = \cos^{-1}(3\sqrt{102}/34) \approx 0.47$.

41. (c) Let (a,b,c) denote the direction numbers of the desired line. This line must be perpendicular to both normal vectors, so $(a,b,c) \cdot (1,1,2) = 0$ and $(a,b,c) \cdot (2,3,2) = 0$. These give $a + b + 2c = 0$ and $2a + 3b + 2c = 0$, so $-2c = a + b = 2a + 3b$. Hence $b = -a/2$. Substitute for b in the first equation to get $a/2 + 2c = 0$, meaning $c = -a/4$. The scale of the direction numbers is arbitrary, so let $a = -4$. Then $b = 2$ and $c = 1$, and the line is $(x - 1)/(-4) = (y - 1)/2 = (z - 2)$. In vector form, it is $(1,1,2) + t(-4,2,1)$.

45. We want to minimize $C = \ell_1(a + bD_1) + \ell_2(a + bD_2)$, where ℓ_1, ℓ_2, D_1, and D_2 are the variables. Using the method of Lagrange multipliers, we let $f(\ell_1, \ell_2, D_1, D_2, \lambda_1, \lambda_2) = [\ell_1(a + bD_1) + \ell_2(a + bD_2) - C] - \lambda_1(\ell_1 + \ell_2 - \ell) - \lambda_2[kQ^m(\ell_1/D_1^{m1} + \ell_2/D_2^{m2}) - h]$. Taking partials, we get $\partial f/\partial \ell_1 = a + bD_1 - \lambda_1 - \lambda_2 kQ^m/D_1^{m1}$; $\partial f/\partial \ell_2 = a + bD_2 - \lambda_1 - \lambda_2 kQ^m/D_2^{m2}$; $\partial f/\partial D_1 = \ell_1 b + m_1 \lambda_2 kQ^m \ell_1/D_1^{m1+1}$; $\partial f/\partial D_2 = \ell_2 b + m_2 \lambda_2 kQ^m \ell_2/D_2^{m2+1}$; $\partial f/\partial \lambda_1 = \ell_1 + \ell_2 - \ell$; and $\partial f/\partial \lambda_2 = kQ^m(\ell_1/D_1^{m1} + \ell_2/D_2^{m2})$. To minimize C, we set all of the partials equal to zero. From $\partial f/\partial D_1 = 0$ and $\partial f/\partial D_2 = 0$, we get the simplified equations $b = -m_1 \lambda_2 kQ^m/D_1^{m1+1} = -m_2 \lambda_2 kQ^m/D_2^{m2+1}$. If $m_1 = m_2$, then $D_1 = D_2$.

   If $m_1 \neq m_2$, we rearrange the last equation to get $1 = (m_1/m_2) \times (D_2^{m2+1}/D_1^{m1+1})$ or $m_1 D_2^{m2+1} = m_2 D_1^{m1+1}$. From $\partial f/\partial \ell_1 = 0$ and $\partial f/\partial \ell_2 = 0$, we simplify to $bD_2 = bD_1 - \lambda_2 kQ^m/D_1^{m1} + \lambda_2 kQ^m/D_2^{m2}$. Putting everything over the common denominator $D_1^{m1}D_2^{m2}$, we get $bD_1^{m1}D_2^{m2+1} = bD_2^{m2}D_1^{m1+1}$. Division by $bD_1^{m1}D_2^{m2}$ yields $D_2 = D_1$. Thus, $D_1 = D_2$ independent of m_1 and m_2.

49. We look for the maximum on the circle of radius 2 centered at the origin. The maximum is about 570 in the first quadrant.

53. (a) $dy/dx = -(\partial F/\partial x)/(\partial F/\partial y) = -(2x)/(3y^2 + e^y)$.

(b) By the chain rule, we have $(\partial F_1/\partial x)(dx/dx) + (\partial F_1/\partial y_1)(dy_1/dx) + (\partial F_1/\partial y_2)(dy_2/dx) = 0$ and $(\partial F_2/\partial x)(dx/dx) + (\partial F_2/\partial y_1)(dy_1/dx) + (\partial F_2/\partial y_2)(dy_2/dx) = 0$. Using the fact that $dx/dx = 1$, multiply the first equation by $\partial F_2/\partial y_2$ and the second by $-\partial F_1/\partial y_2$, and add. This yields $(\partial F_2/\partial y_2)(\partial F_1/\partial x) - (\partial F_1/\partial y_2)(\partial F_2/\partial x) + [(\partial F_2/\partial y_2)(\partial F_1/\partial y_1) - (\partial F_2/\partial y_1)(\partial F_1/\partial y_2)](dy_1/dx)$. Rearrangement gives $dy_1/dx = [(\partial F_1/\partial y_2)(\partial F_2/\partial x) - (\partial F_2/\partial y_2)(\partial F_1/\partial x)]/[(\partial F_2/\partial y_2)(\partial F_1/\partial y_1) - (\partial F_2/\partial y_1)(\partial F_1/\partial y_2)]$.

Similarly, multiply the first equation by $\partial F_2/\partial y_1$ and the second by $-\partial F_1/\partial y_1$. This yields $dy_2/dx = [(\partial F_1/\partial y_1)(\partial F_2/\partial dx) - (\partial F_1/\partial x)(\partial F_2/\partial y_1)]/[(\partial F_1/\partial y_2)(\partial F_2/\partial y_1) - (\partial F_2/\partial y_2)(\partial F_1/\partial y_1)]$.

(c) Let $F_1 = x^2 + y_1^2 - \cos x$ and $F_2 = x^2 - y_2^2 - \sin x$. Then $\partial F_1/\partial x = 2x + \sin x$, $\partial F_1/\partial y_1 = 2y_1$, $\partial F_1/\partial y_2 = 0$, $\partial F_2/\partial x = 2x - \cos x$, $\partial F_2/\partial y_1 = 0$, and $\partial F_2/\partial y_2 = 2y_2$. Thus, $dy_1/dx = [0 - 2y_2(2x + \sin x)]/[4y_1y_2 - 0] = -(2x + \sin x)/2y_1$ and $dy_2/dx = [2y_1(2x - \cos x) - 0]/[0 - 4y_1y_2] = (\cos x - 2x)/y_2$.

57. (a) $P_y = 2x^2y \exp(xy^2) + (x^2y^2 + 2x)2xy \exp(xy^2) = (2x^3y^3 + 6x^2y) \times \exp(xy^2)$ and $Q_x = 6x^2y \exp(xy^2) + (2x^3y)y^2 \exp(xy^2) = P_y$; therefore, a function exists. Integrate $f_x = (x^2y^2 + 2x) \exp(xy^2)$ by parts with $u = x^2y^2 + 2x$ and $v = (1/y^2) \exp(xy^2)$ to get $(x^2 + 2x/y^2) \exp(xy^2) - \int(2x + 2/y^2) \exp(xy^2)dx$. Use parts again to get $(x^2 + 2x/y^2) \exp(xy^2) - (2x/y^2 + 2/y^4) \exp(xy^2) + \int(2/y^2) \exp(xy^2) dx = x^2\exp(xy^2) + g(y)$ for some $g(y)$. Integrate f_y by substituting $u = xy^2$ to get $\int 2x^3y \exp(xy^2)dy = \int x^2 e^u du = x^2 e^u + h(x) = x^2\exp(xy^2) + h(x)$ for some $h(x)$. Clearly, $g(y) = h(x) = $ constant, so $f(x,y) = x^2\exp(xy^2) + C$ is the solution.

57. (b) $P_y = (2xy + 3xy^2)e^{xy^3}$ and $Q_x = 2y(3x^2 + y^3)e^{xy^3}$. Since $P_y \neq$

Q_x , no such function exists.

   (c) Integrate $f_x = 2x/(1 + x^2 + y^2)$ to get $f(x,y) = \ln(1 + x^2 + y^2) +$

$g(y)$, for some $g(y)$. Integrate $f_y = 2y/(1 + x^2 + y^2) =$

$\ln(1 + x^2 + y^2) + h(x)$, for some $h(x)$. Clearly, $h(x) = g(y) =$

constant , so $f(x,y) = \ln(1 + x^2 + y^2) + C$ is the function.

   (d) Let $A = (1 + x^2 + y^2)$. Then $P_y = [2A - 4y^2]/A^2$ and $Q_x =$

$[2A - 4x^2]/A^2$. Since $P_y \neq Q_x$, there is no such function.

61. (a) $f_x = -2xy^3/(x^2 + y^2)^2$ if $(x,y) \neq (0,0)$. $f_y = [3y^2(x^2 + y^2) -$

$2y^4]/(x^2 + y^2)^2 = y^2(3x^2 + y^2)/(x^2 + y^2)^2$ if $(x,y) \neq (0,0)$.

In order to deal with $f_x(0,0)$ and $f_y(0,0)$, approach the origin

along the line $y = 0$ for f_x and along the line $x = 0$ for f_y .

Hence, $f_x(0,0) = \lim_{x \to 0}[-2x \cdot 0/x^4] = \lim_{x \to 0}(0/x^4) = 0$; and $f_y(0,0) =$

$\lim_{y \to 0}[y^2(0 + y^2)/y^4] = \lim_{y \to 0}(y^4/y^4) = 1$.

   (b) We have $x = r \cos \theta$ and $y = r \sin \theta$, so $\partial x/\partial r = \cos \theta$ and

$\partial y/\partial r = \sin \theta$. Compute $(\partial/\partial r)f(r \cos \theta, y \sin \theta) = (\partial/\partial r)f(x,y) =$

$f_x(\partial x/\partial r) + f_y(\partial y/\partial r) = -2xy^3 \cos \theta/(x^2 + y^2)^2 + y^2(3x^2 + y^2)/(x^2 +$

$y^2)^2 = -2r^4 \cos^2\theta \sin^3\theta + 3r^4 \sin^2\theta \cos^2\theta + r^4 \cos^2\theta/r^4 =$

$\cos^2\theta(-2 \sin^3\theta + 3 \sin^2\theta + 1)$. This quantity is defined for all

θ (and is 1 when $\theta = 0$).

   (c) Let D_θ denote the directional derivative in the direction θ at

the origin. Suppose $D_\theta = \underline{\nabla}f \cdot (\cos \theta, \sin \theta) = f_x \cos \theta + f_y \sin \theta =$

$\sin \theta$. However, this disagrees with (b). This result does not

contradict the chain rule because the partials are not continuous

at $(0,0)$.

TEST FOR CHAPTER 16

1. True or false.

    (a) If a two-variable function f is defined throughout the disk
         $x^2 + y^2 \leqslant 1$, then f must have a minimum and a maximum, even
         if a saddle point occurs at the origin.

    (b) If F is a function of x and y such that $y = f(x)$, then
         $dy/dx = (\partial F/\partial x)/(\partial F/\partial y)$.

    (c) The gradient is a directional derivative.

    (d) If f is a differentiable function and a local minimum occurs
         at (x_0,y_0) , then $f_x(x_0,y_0) = f_y(x_0,y_0) = 0$.

    (e) If the gradient $\underline{\nabla} f(x_0,y_0)$ is nonzero, it is normal to the surface
         $z = f(x,y)$ at (x_0,y_0,z_0) .

2.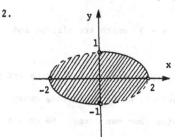
    Locate the minimum and maximum (if they exist)
    for $f(x,y) = xy^2 + x^2 - 1$ over the region
    shown at the left. The boundary is the ellipse
    $x^2/4 + y^2 = 1$, but it is part of the region
    only in the first and third quadrants.

3. (a) Find the maximum of xyz subject to the constraints $x + y + z = 32$
         and $x - y + z = 0$.

    (b) Explain what this means geometrically.

4. Classify the critical points of $f(x,y) = x^3 - 3xy + y^3 - 5$.

5. (a) What is the relationship between the gradient and the tangent plane?

    (b) Compute the tangent plane to the surface $3x^3 - 9xy - 3z^2y +$
         $z = 26$ at the point $(1,-1,2)$.

6.   Let $f(x,y,z) = e^{xy} + e^{xz} + e^{2yz}$.

   (a)   Compute the directional derivative of f at $(1,1,1)$ in the
   direction $\underline{i} + 2\underline{j} - 2\underline{k}$.

   (b)   Explain what the answer in part (a) tells you.

   (c)   In what direction is f increasing most rapidly at $(1,1,1)$?

7.   A rectangular box has sides with length x , y , and $2x$. Given
   100 square units of surface area, what is the maximum volume that can be
   held by the box?   Use Lagrange multipliers.

8.   Let $z = F(x,y) = \cos(x^2 y^2) + 2xy$.

   (a)   How is dy/dx related to F_x and F_y ?

   (b)   Use the formula in (a) to compute dy/dx .

   (c)   Compute the directional derivative of F at $(1, \sqrt{\pi})$ in the direction
   of $\underline{i} - \underline{j}$.

9.   Find the points on the plane $5x + 2y - z = 3$ which are closest and
   farthest from the point $(3,0,2)$.

10. Your worst enemy has cursed you with $f(x,y)$ years of bad luck after you
   "accidentally" stepped on his foot. However, being a sporting human
   being, he offers you a chance to determine your own curse. He owns a
   device which randomly selects a value for $f(x,y) = x^2 + \sin y$ over
   the rectangle $1 \leqslant x \leqslant 3$ and $\pi/2 \leqslant y \leqslant \pi$. Your worst enemy offers
   you a chance to push the button to determine your curse. Find the minimum
   and maximum number of years your worst enemy plans to curse you with
   bad luck.

ANSWERS TO CHAPTER TEST

1. (a) True

    (b) False; negative sign is missing.

    (c) False; the gradient is a vector; the directional derivative is a scalar.

    (d) True

    (e) True

2. Minimum at $(0,0)$; no maximum.

3. (a) 1024 at $(8,16,8)$.

    (b) It is the maximum of xyz on the line which forms the intersection

    of the planes $x + y + z = 32$ and $x - y + z = 0$.

4. Saddle point at $(0,0)$; local minimum at $(1,1)$.

5. (a) The tangent plane is $\nabla f(\underline{r}_0) \cdot (\underline{r} - \underline{r}_0) = 0$ where $\underline{r} = x\underline{i} + y\underline{j} + z\underline{k}$.

    (b) $18x - 21y + 13z = 65$

6. (a) $2e/3$

    (b) If one moves one unit in direction $\underline{i} + 2\underline{j} - 2\underline{k}$, f should increase

    $2e/3$ units.

    (c) $2e\underline{i} + (e + 2e^2)\underline{j} + (e + 2e^2)\underline{k}$

7. $x = 5/\sqrt{3}$; $y = 20/3\sqrt{3}$.

8. (a) $dy/dx = -F_x/F_y$

    (b) $[1 - xy \sin(x^2y^2)]/[xy \sin(x^2y^2) - 1]$

    (c) $\sqrt{2\pi} - \sqrt{2}$

9. Closest: $(4/3,-2/3,43/15)$. Farthest: none.

10. Minimum $= 1$; maximum $= 10$.

CHAPTER 17

MULTIPLE INTEGRATION

17.1 The Double Integral and Iterated Integral

PREREQUISITES

1. Recall the definition of the integral in one variable (Section 4.3).
2. Recall the basic rules of integration (Chapter 7).

PREREQUISITE QUIZ

1. Define an upper sum.
2. What is the relationship between lower sums, upper sums, and the integral?
3. Compute the following integrals:
    (a) $\int (x^3 + 2 + 1/x + \cos x)dx$
    (b) $\int x \sin x \, dx$
    (c) $\int (e^t + \sqrt{2t + 2})dt$
    (d) $\int_1^2 x^2 dx$

GOALS

1. Be able to define the double integral in terms of lower sums and upper sums.
2. Be able to evaluate double integrals over rectangular regions by using iterated integrals.

STUDY HINTS

1. Notation. [a,b] × [c,d] describes the rectangle $a \leq x \leq b$ and $c \leq y \leq d$. As with the single variable notation, the use of parentheses rather than brackets means that the boundary is not included in the rectangle. A double integral over a rectangle is written as $\int_c^d \int_a^b f(x,y) dx \, dy$. The inner differential dx is associated with the inner limits a and b . Similarly, the outer differential is associated with the outer limits. Sometimes, it is written $\int\int_D f(x,y) dx \, dy$ to indicate integration over the region D = [a,b] × [c,d] .

2. Dummy variables. Just as the variable x was commonly used in one variable calculus, x and y are commonly used in two variable calculus. However, any other letter may also be used. For example, $\int_0^1 \int_2^3 f(x,y) dx \, dy = \int_0^1 \int_2^3 g(u,v) \, du \, dv$ if f(x,y) = g(u,v) .

3. Geometry. In one-variable integration, the integral represented the area under the curve, provided the integrand is nonnegative. With two variables, another dimension is added. The geometric interpretation is the volume under the surface, provided the integrand is nonnegative.

4. Definition. Again, the double integral is that unique number which falls between all upper and all lower sums.

5. Properties. The properties of double integrals are analogous to those of ordinary integrals. If you understand each statement, there should be no need to memorize them.

6. Iterated integrals. The iterated integral is simply the double integral written with brackets: $\int_c^d [\int_a^b f(x,y) dx] dy$ or $\int_a^b [\int_c^d f(x,y) dy] dx$. The brackets indicate that you evaluate the inner integral first and then the outer one. Notice that the order of integration doesn't matter.

7. **Computing double integrals.** This is just a reverse of partial differ-
    entiation. All variables are held constant except for the one upon
    which the integration is performed.

SOLUTIONS TO EVERY OTHER ODD EXERCISE

1. The integral over each subrectangle is the value of g times the area:
    $\iint g(x,y)dx\,dy = -2 \times 3 + 0 \times 1 + 1 \times 9 + 3 \times 3 = 12$.

5. $\int_0^3\int_0^2 x^3y\,dx\,dy = \int_0^3(x^4y/4)\big|_{x=0}^2 dy = \int_0^3 4y\,dy = 2y^2\big|_0^3 = 18$.

9. $\int_{-1}^1\int_0^1 ye^x dy\,dx = \int_{-1}^1 (y^2e^x/2)\big|_{y=0}^1 dx = (1/2)\int_{-1}^1 e^x dx = (1/2)e^x\big|_{-1}^1 = e/2 -$
    $1/2e$.

13. The integral is $\int_{-2}^2\int_0^1(x^2 + 2xy - y\sqrt{x})dx\,dy = \int_{-2}^2(x^3/3 + x^2y -$
    $(2/3)yx^{3/2})\big|_{x=0}^1 dy = \int_{-2}^2(1/3 - y/3)dy = (y/3 - y^2/6)\big|_{-2}^2 = 4/3$.

17. $\int_2^4\int_{-1}^1 x(1 + y)dx\,dy = \int_2^4(1 + y)dy \cdot \int_{-1}^1 x\,dx = (y + y^2/2)\big|_2^4 \cdot (x^2/2)\big|_{-1}^1 =$
    $(8)(0) = 0$. This is what we expected from part (b) of Exercise 3.

21. The mass is the double integral over the density, as in Example 8. Let
    the center of the plate be located at the origin. Then $r^2 = x^2 + y^2$
    and the mass is $\int_{-1/2}^{1/2}\int_{-1/2}^{1/2}(4 + x^2 + y^2)dx\,dy =$
    $\int_{-1/2}^{1/2}(4x + x^3/3 + xy^2)\big|_{x=-1/2}^{1/2} dy = \int_{-1/2}^{1/2}(49/12 + y^2)dy =$
    $(49y/12 + y^3/3)\big|_{-1/2}^{1/2} = 25/6$ grams.

25. (a) At time t on day T , $[1 - \sin^2\alpha \cos^2(2\pi T/365)]^{1/2}\cos(2\pi t/24)$
        and $\sin\alpha \cos(2\pi T/365)$ are both constants. Denote these two
        constants by A and B , respectively. Thus, $I = A\cos\ell +$
        $B\sin\ell$; ℓ is in degrees and we would like to integrate in the
        same units. Colorado is 6° longitude wide, which is equivalent
        to 660 km or 8° latitude. The total solar energy is the double
        integral $\iint_D I\,dx\,d\ell$, where x is the east-west distance. Thus,
        the integrated solar energy is $8\int_{33}^{41}(A\cos\ell + B\sin\ell)d\ell =$
        $8(A\sin\ell - B\cos\ell)\big|_{33}^{41} = 0.88A + 0.67B$.

25. (b) Integrating the result of part (a) from t_1 to t_2 is the total

    solar energy received in the state between the times t_1 and t_2

    on day T .

SECTION QUIZ

1. (a) Compute $\int_0^2 \int_1^4 (x^3 y + xy^2) \, dy \, dx$.

   (b) Explain the geometric interpretation of this integral.

2. The integral $\int_a^b \int_c^d xy \, dy \, dx$ is $(\int_a^b x \, dx)(\int_c^d y \, dy) = (x^2/2)\big|_a^b \cdot (y^2/2)\big|_c^d$.

   (a) Under what conditions can the above method of multiplying one-variable

    integrals be used to evaluate double integrals?

   (b) Use the technique described in part (a) to integrate

    $\int_a^b \int_c^d (x - 1)(y^2 + y) \, dy \, dx$ or explain why the method is inappropriate.

3. Find a formula for the volume between two surfaces $z = f(x,y)$ and

   $z = g(x,y)$ over the rectangle $[a,b] \times [c,d]$, assuming $f(x,y) > g(x,y)$

4. 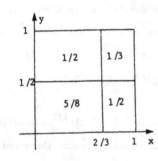   A monster has begun to devour downtown

    Cleveland. The monster began its rampage

    by quickly eating up the block $0 \leqslant x \leqslant 1$,

    $0 \leqslant y \leqslant 1$. The four buildings on the block

    have height, width, and length as diagrammed.

    For example, one building has width $1/2$,

    length $2/3$, and height $5/8$.

   (a) If $f(x,y)$ is the height of the buildings, write the volume of the

    buildings on the block as a double integral.

   (b) How many cubic blocks did the monster consume when it devoured

    the first block of buildings?

ANSWERS TO PREREQUISITE QUIZ

1. In one-variable calculus, an upper sum is any number which is greater than the area under the graph of f .

2. The integral is precisely that number which lies between all upper sums and all lower sums.

3. (a) $x^4/4 + 2x + \ln |x| + \sin x + C$

    (b) $-x \cos x + \sin x + C$

    (c) $e^t + (2t + 2)^{3/2}/3 + C$

    (d) $7/3$

ANSWERS TO SECTION QUIZ

1. (a) 72

    (b) The volume between $f(x,y) = x^3 y + xy^2$ and the xy-plane over the rectangle $[0,2] \times [1,4]$ is 72 .

2. (a) When the integrand can be factored in functions of y only and x only, and the limits of integration are constant.

    (b) $\int_a^b (x - 1)dx \cdot \int_c^d (y^2 + y)dy = [(b^2 - a^2)/2 - (b - a)][(d^3 - c^3)/3 - (d^2 - c^2)/2]$

3. $\int_a^b \int_c^d [f(x,y) - g(x,y)] dy\, dx$

4. (a) $\int_0^1 \int_0^1 f(x,y)dx\, dy$

    (b) $37/72$

17.2 The Double Integral Over General Regions

PREREQUISITES

1.   Recall how to integrate by using iterated integrals (Section 17.1).

PREREQUISITE QUIZ

1.   Compute the following integrals:

(a) $\int_0^3 \int_{-2}^0 xy^2 \, dx \, dy$

(b) $\int_0^3 \int_{-2}^0 xy^2 \, dy \, dx$

(c) $\int_0^\pi \int_0^1 u \cos v \, du \, dv$

GOALS

1.   Be able to integrate a double integral over a general planar region.

2.   Be able to change the order of integration in double integrals and identify the corresponding region.

STUDY HINTS

1.   Regions for theory. The notations D* , D** , f* , f** are intro-
     duced so that the transition can be made from rectangular regions to
     general regions. You won't need to worry about these concepts for
     solving most of the problems.

2.   Region types. Type 1 regions are enclosed by two curves which are graphs
     of functions of x , and the lines x = a and x = b . Type 2 regions
     are bounded by two curves which are graphs of functions of y , and the
     lines y = a and y = b . See Figs. 17.2.3 and 17.2.4. Recognizing the
     region type will help you set up an integral. Naming the regions type 1
     and type 2 is most important for discussion purposes. Your primary con-
     cern should be learning to perform multiple integration, not learning
     how to name the regions.

3. <u>Simplifying complicated regions</u>. Any planar region may be broken up

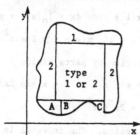

into regions of type 1 and type 2. For
example, the region at the left is divided
into seven pieces, each of which is either
type 1 or type 2. **Fig. 17.2.5 and regions A ,
B , and C in the figure at the left show how**
some regions may be both type 1 <u>and</u> type 2.

4. <u>Integration method</u>. For type 1 regions, it is best to integrate in y
and then in x . For type 2 regions, integrate first in x , then y .

5. <u>Choosing integration limits</u>. When you perform multiple integration, be
sure that the limits of integration do not include any previously integrated variables. In particular, no variable should appear in the limits
of the outer most integral. For example, the integral of $(x + y)^2$ over
the region D: $0 \leqslant y \leqslant x^2, 0 \leqslant x \leqslant 1$, should be written as
$\int_0^1 \int_0^{x^2} (x + y)^2 dy\, dx$, <u>not</u> $\int_0^{x^2} \int_0^1 (x + y)^2 dx\, dy$. Sketching the region
often helps in choosing your limits.

SOLUTIONS TO EVERY OTHER ODD EXERCISE

1.

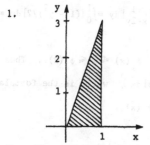

This is both a type 1 and a type 2 region. It
can be described by x in [0,1] and $\phi_1(x) =$
$0 \leqslant y \leqslant 3x = \phi_2(x)$. On the other hand, it can
be described by y in [0,3] and $\psi_1(y) =$
$y/3 \leqslant x \leqslant 1 = \psi_2(y)$.

5. The region D is described by $0 \leqslant x \leqslant 1$ and $0 \leqslant y \leqslant x$. Thus,
$\iint_D (x + y)^2 dx\, dy = \int_0^1 \int_0^x (x + y)^2 dy\, dx = \int_0^1 [(x + y)^3/3] \,|_{y=0}^x \, dx =$
$\int_0^1 (8x^3/3 - x^3/3) dx = \int_0^1 (7x^3/3) dx = (7x^4/12)\,|_0^1 = 7/12$.

9.

This region is type 1.

$\int_0^\pi \int_{\sin x}^{3 \sin x} x(1 + y)dy\, dx = \int_0^\pi [x(1 + y)^2/$

$2] \Big|_{y=\sin x}^{3 \sin x} dx = 2\int_0^\pi x(\sin x + 2 \sin^2 x)dx$.

Then, integration by parts yields

$2[-x \cos x + \sin x + x^2/2 - x \sin 2x/2 -$

$\cos 2x/2] \Big|_0^\pi = 2\pi + \pi^2$.

13.

This is a type 1 region. The integral is

$\int_0^1 \int_0^{x^2} (x^2 + xy - y^2)dy\, dx = \int_0^1 [(x^2 y + xy^2/2 -$

$y^3/3) \Big|_{y=0}^{x^2}]dx = \int_0^1 (x^4 + x^5/2 - x^6/3)dx =$

$(x^5/5 + x^6/12 - x^7/21) \Big|_0^1 = 1/5 + 1/12 - 1/21 =$

$33/140$.

17.

When the order of integration is interchanged,

we have $0 \leqslant y \leqslant 1$ and $0 \leqslant x \leqslant y$. Thus, the

integral is $\int_0^1 \int_0^y xy\, dx\, dy = \int_0^1 y(x^2/2) \Big|_{x=0}^y dy =$

$\int_0^1 (y^3/2)dy = (y^4/8) \Big|_0^1 = 1/8$.

21.

D is described by $0 \leqslant y \leqslant 1$ and

$2y \leqslant x \leqslant y + 1$, so the integral is

$\int_0^1 \int_{2y}^{y+1} (x - y)\, dx\, dy =$

$\int_0^1 [(x^2/2 - xy) \Big|_{x=2y}^{y+1}]dy = \int_0^1 [(1 - y^2)/2]dy =$

$(1/2)(y - y^3/3) \Big|_0^1 = (1/2)(2/3) = 1/3$.

25. For a type 1 region, we have $a \leqslant x \leqslant b$ and $\phi_1(x) \leqslant y \leqslant \phi_2(x)$. Thus,

$\iint_D dx\, dy = \int_a^b \int_{\phi_1(x)}^{\phi_2(x)} dy\, dx = \int_a^b [\phi_2(x) - \phi_1(x)]dx$, which is the formula

for the area between the curves $\phi_2(x)$ and $\phi_1(x)$.

SECTION QUIZ

1.

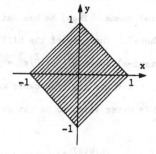

Integrate $x^2 + y^2$ over the square region shown in the figure.

2.

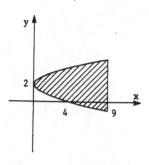

Integrate xy over the region shown in the figure.

3.

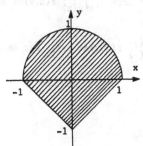

Integrate $2x + y$ over the parabolic region shown in the figure.

4.   A region in the first quadrant is described by $0 \leq y \leq 2 - 2x$ and $0 \leq x \leq 1$. We wish to integrate $f(x,y) = x$ over this region.

(a)   Sketch the region.

(b)   Explain what is wrong with $\int_0^{2-2x} \int_0^1 x \, dx \, dy$.

(c)   Compute the double integral.

5.   The owner of a new amusement park wants his customers to begin enjoying
     themselves as soon as they leave their hotel rooms. Thus, he has built
     a water slide along the back wall of the hotel. The base of the hill
     on which the water slide is built is the semi-ellipse $x^2 + 4y^2/\pi^2 = 1$,
     $x \leqslant 0$. The height of the hill is $f(x,y) = x + \cos y$. The owner wants
     to know how much dirt he bought to build his water slide so he can deduct
     it from his income tax.

     (a)   Express the volume of the hill in the form $\int_a^b \int_{g(x)}^{h(x)} f(x,y)dy\ dx$,
           but do not evaluate.

     (b)   Express the volume in the form $\int_c^d \int_{v(y)}^{u(y)} f(x,y)dx\ dy$, but do not
           evaluate.

ANSWERS TO PREREQUISITE QUIZ

1.   (a) -18

     (b)   12

     (c)   0

ANSWERS TO SECTION QUIZ

1.   2/3

2.   0

3.   460.8

4.   (a)

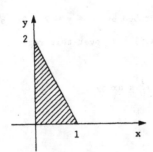

4. (b) The outermost limits contain a variable which has already been
        integrated.

    (c) 1/3

5. (a) $\int_{-1}^{0} \int_{-(\pi/2)\sqrt{1-x^2}}^{(\pi/2)\sqrt{1-x^2}} (x + \cos y) dy\ dx$

    (b) $\int_{-\pi/2}^{\pi/2} \int_{-\sqrt{1-4y^2/\pi^2}}^{0} (x + \cos y) dx\ dy$

17.3 Applications of the Double Integral

PREREQUISITES

1.   Recall how to compute a double integral (Sections 17.1 and 17.2).

2.   Recall how to compute the average of a function of one variable
      (Section 9.3).

3.   Recall how to find the center of mass for functions of one variable
      (Section 9.4).

4.   Recall how to find surface areas of graphs of functions of one variable
      revolved about an axis (Section 10.3).

PREREQUISITE QUIZ

1.   What is the average value of $f(x) = x^2$ over the interval $[0,1]$?

2.   If $f(x) \geqslant 0$ on $[a,b]$, state the two integration formulas used to
      compute $\bar{x}$ and $\bar{y}$, the center of mass of the region under the graph
      of $f(x)$ on $[a,b]$.

3.   If the graph of $f(x) \geqslant 0$ on $[a,b]$ is revolved around the x-axis,
      what formula is used to compute its surface area of revolution?

4.   If the graph of $f(x) \geqslant 0$ on $[a,b]$ is revolved around the y-axis,
      what formula is used to compute its surface area of revolution?

5.   Compute $\int_0^1 \int_0^x 3 \, dy \, dx$.

GOALS

1.   Be able to compute the volume under a surface or the area of a region
      by using double integrals.

2.   Be able to apply double integration to compute an average, a center of
      mass, or a surface area.

STUDY HINTS

1. Volume and area. Recall that the geometric interpretation of the double

    integral is the volume under the surface $z = f(x,y)$. If the height

    is $f(x,y) = 1$, then the integral represents the area of the region

    D: integral = volume = (area of base)(height) = area of D , since

    the height is 1 .

2. Example 1 factorization. You were asked to explain the factorization

    at the conclusion of Example 1. Here is one: $10^4/3^4 \cdot 2^4 + 10 \cdot 5^3/3^2 \cdot 2^3 -$

    $5^4/3 \cdot 2^4 = 2^4 \cdot 5^4/3^4 \cdot 2^4 + 2 \cdot 5^4/3^2 \cdot 2^3 - 5^4/3 \cdot 2^4 = (2^4 \cdot 5^4 + 2^2 \cdot 3^2 \cdot 5^4 -$

    $3^3 \cdot 5^4)/3^4 \cdot 2^4 = (16 + 36 - 27)5^4/3^4 \cdot 2^4 = 25(5^4)/3^4 \cdot 2^4 = 5^6/3^4 \cdot 2^4$.

    Sometimes manipulations like this can be easier than getting a big

    fraction as in the text.

3. Average. In one-variable calculus, the average of a function was the

    integral divided by the length of the interval of integration. Now that

    we are dealing with an integral over a region, we divide by the area, so

    the average is the integral divided by the area of the region of

    integration.

4. Center of mass. Since mass is density times area, we have, for constant

    density, mass $= \rho \iint_D dx\, dy$. If density is a function of x and y ,

    mass becomes $\iint_D \rho(x,y)dx\, dy$. By applying the infinitesimal argument

    of Section 9.4, we get $\bar{x} = \iint_D x\rho(x,y)dx\, dy/\iint_D \rho(x,y)dx\, dy$ and $\bar{y} =$

    $\iint_D y\rho(x,y)dx\, dy/\iint_D \rho(x,y)dx\, dy$.

5. Surface area. The best thing to do is to memorize the formula in the box

    on p. 857. The integrand resembles the arc length integrand in one vari-

    able except that an extra differential term is under the radical.

SOLUTIONS TO EVERY OTHER ODD EXERCISE

1.   The volume is $\int_c^d \int_a^b f(x,y)\,dx\,dy$. In this case, it is

$\int_\pi^{3\pi} \int_0^2 (x \sin y + 3)\,dx\,dy = \int_\pi^{3\pi} [(x^2 \sin y/2 + 3x)|_{x=0}^2]\,dy =$

$\int_\pi^{3\pi} (2 \sin y + 6)\,dy = (-2 \cos y + 6y)|_\pi^{3\pi} = 12\pi$.

5.

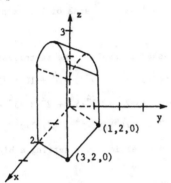

The parallelogram is described by

$0 \leqslant y \leqslant 1$ and $y/2 \leqslant x \leqslant y/2 + 2$,

so the volume is $\int_0^2 \int_{y/2}^{y/2+2} (1 +$

$\sin(\pi y/2) + x)\,dx\,dy = \int_0^2 [(x +$

$x \sin(\pi y/2) + x^2/2)|_{x=y/2}^{y/2+2}]\,dy =$

$\int_0^2 (4 + 2\sin(\pi y/2) + y)\,dy =$

$[4y - (4/\pi)\cos(\pi y/2) + y^2/2]|_0^2 =$

$10 + 8/\pi$.

9.

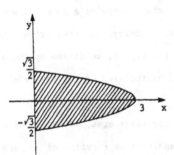

The problem is to find the volume

between the surface $z = x^3 y$ and

the xy-plane over the region shown

in the figure. Due to the symmetry,

we will only compute the volume above

the xy-plane and then, multiply by 2 .

The region on the xy-plane is described

by $0 \leqslant y \leqslant \sqrt{3 - x}/2$ and $0 \leqslant x \leqslant 3$.   Therefore, the half-volume is

$\int_0^3 \int_0^{\sqrt{3-x}/2} x^3 y\,dy\,dx = \int_0^3 (x^3 y^2/2)\Big|_{y=0}^{\sqrt{3-x}/2}\,dx = (1/8)\int_0^3 (3x^3 - x^4)\,dx =$

$243/160$.   Thus, the entire volume is $243/80$.

13.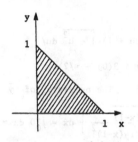

The average value over a region is

$\iint_D f(x,y)dx\,dy / \iint_D dx\,dy$. The region

D is described by $0 \leqslant x \leqslant 1$ and

$0 \leqslant y \leqslant 1 - x$, so the numerator is

$\int_0^1\int_0^{1-x} e^{x+y}\,dy\,dx = \int_0^1 (e^{x+y}\big|_{y=0}^{1-x})\,dx =$

$\int_0^1 (e - e^x)dx = (ex - e^x)\big|_0^1 = 1$. The

denominator is $\int_0^1\int_0^{1-x} dy\,dx = \int_0^1(1 - x)dx = (x - x^2/2)\big|_0^1 = 1/2$, so

the average value is $1/(1/2) = 2$.

17. The formula for the average value over a region is $\iint_D f(x,y)dx\,dy /$

$\iint_D dx\,dy$. The numerator is $\int_0^a\int_0^a (x^2 + y^2)dx\,dy = \int_0^a [(x^3/3 +$

$xy^2)\big|_{x=0}^a]\,dy = \int_0^a (a^3/3 + ay^2)dy = (a^3y/3 + ay^3/3)\big|_0^a = 2a^4/3$. The

denominator is $\int_0^a\int_0^a dx\,dy = \int_0^a a\,dy = a^2$, so the average value is

$(2a^4/3)/a^2 = 2a^2/3$.

21. The center of mass is given by $\bar{x} = \iint_D x\rho(x,y)dx\,dy / \iint_D \rho(x,y)dx\,dy$ and

$\bar{y} = \iint_D y\rho(x,y)dx\,dy / \iint_D \rho(x,y)dx\,dy$. The disk is described by

$-\sqrt{1 - (x - 1)^2} \leqslant y \leqslant \sqrt{1 - (x - 1)^2}$ and $0 \leqslant x \leqslant 2$. When $\rho(x,y) =$

x^2 , the following integrations will be used: $\int_{-1}^1 u^2\sqrt{1 - u^2}\,du =$

$[(u/8)(2u^2 - 1)\sqrt{1 - u^2} + (1/8) \sin^{-1}u]\big|_{-1}^1 = \pi/8$; $\int_{-1}^1 u\sqrt{1 - u^2}\,du =$

$(-1/3)(1 - u^2)^{3/2}\big|_{-1}^1 = 0$; $\int_{-1}^1\sqrt{1 - u^2}\,du = [(u/2)\sqrt{1 - u^2} +$

$(1/2) \sin^{-1}u]\big|_{-1}^1 = \pi/2$. These formulas came from the integration table

(numbers 38, 51, and 52). Also, let $v = 1 - u^2$, so $u^2 = 1 - v$, and

$\int_{-1}^1 u^3\sqrt{1 - u^2}\,du = -\int_0^0(1 - v)\sqrt{v}(dv/2) = 0$. Therefore,

$\bar{x} = \int_0^2\int_{-\sqrt{1-(x-1)^2}}^{\sqrt{1-(x-1)^2}} x^2(x)\,dy\,dx / \int_0^2\int_{-\sqrt{1-(x-1)^2}}^{\sqrt{1-(x-1)^2}} x^2\,dy\,dx =$

$\int_0^2\left(yx^3\big|_{y=-\sqrt{1-(x-1)^2}}^{\sqrt{1-(x-1)^2}}\right)dx / \int_0^2\left(yx^2\big|_{y=-\sqrt{1-(x-1)^2}}^{\sqrt{1-(x-1)^2}}\right)dx = 2\int_0^2 x^3\sqrt{1 - (x-1)^2}\,dx /$

$2\int_0^2 x^2\sqrt{1 - (x - 1)^2}\,dx$. Let $u = x - 1$ to get $\int_{-1}^1 (u + 1)^3\sqrt{1 - u^2}\,du /$

21. (continued)

$$\int_{-1}^{1}(u + 1)^2\sqrt{1 - u^2} \; du = \int_{-1}^{1}(u^3 + 3u^2 + 3u + 1)\sqrt{1 - u^2} \; du/$$

$$\int_{-1}^{1}(u^2 + 2u + 1)\sqrt{1 - u^2} \; du = (0 + 3(\pi/8) + 3(0) + \pi/2)/$$

$(\pi/8 + 2(0) + \pi/2) = (7\pi/8)/(5\pi/8) = 7/5$. The numerator of $\bar{y}$ is

$$\int_{0}^{2}\int_{-\sqrt{1-(x-1)^2}}^{\sqrt{1-(x-1)^2}} yx^2 \; dy \; dx = \int_{0}^{2}\left(y^2x^2/2 \Big|_{y=-\sqrt{1-(x-1)^2}}^{\sqrt{1-(x-1)^2}}\right) dx = \int_{0}^{2}0 \; dx = 0 \; .$$

Thus, the center of mass is $(7/5,0)$.

25.

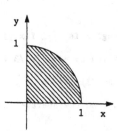

We want the area of $z = \sqrt{x^2 + y^2}$ lying above the region described by $0 \leqslant x \leqslant 1$ and $0 \leqslant y \leqslant \sqrt{1 - x^2}$. $f_x = x/\sqrt{x^2 + y^2}$ and $f_y = y/\sqrt{x^2 + y^2}$. Thus, the surface area is

$$\int_{0}^{1}\int_{0}^{\sqrt{1-x^2}} \sqrt{1 + x^2/(x^2 + y^2) + y^2/(x^2 + y^2)} \, dy \, dx$$

$$\int_{0}^{1}\int_{0}^{\sqrt{1-x^2}} \sqrt{2} \; dy \; dx = \int_{0}^{1}\sqrt{2}\sqrt{1 - x^2} \; dx \; .$$ Using formula 38 of the integration

table, this is $\sqrt{2}[(x/2)\sqrt{1 - x^2} + (1/2) \sin^{-1}x]\Big|_{0}^{1} = \sqrt{2}(\pi/4)$.

29. (a)

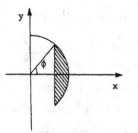

Center the area around the x-axis. Then the spherical area lies above and below the region shown at the left. It can be described by $r \cos \phi \leqslant x \leqslant r$ and since $x^2 + y^2 = r^2$, we have $-\sqrt{r^2 - x^2} \leqslant y \leqslant \sqrt{r^2 - x^2}$. The equation of the sphere is $x^2 + y^2 + z^2 = r^2$, so $z = \pm\sqrt{r^2 - x^2 - y^2}$. For the area lying above the xy-plane, $f_x = -x/\sqrt{r^2 - x^2 - y^2}$ and $f_y = -y/\sqrt{r^2 - x^2 - y^2}$. Thus, the upper area is

$$\int_{r \cos \phi}^{r}\int_{-\sqrt{r^2-x^2}}^{\sqrt{r^2-x^2}} \left(r/\sqrt{r^2 - x^2 - y^2}\right)dy \; dx \; .$$ Factor out $1/\sqrt{r^2 - x^2}$

and let $u = y/\sqrt{r^2 - x^2}$ to get $\int_{r \cos \phi}^{r}\int_{-1}^{1} (r/\sqrt{1 - u^2} \; du \; dx =$

29. (a) (continued)

$\int_r^r \cos\phi^{(r \sin^{-1}u|_{-1}^1)}dx = \pi r \int_r^r \cos\phi \, dx = \pi r^2(1 - \cos\phi)$. Since

there is also a bottom half, the entire area is $2\pi r^2(1 - \cos\phi)$.

(b)

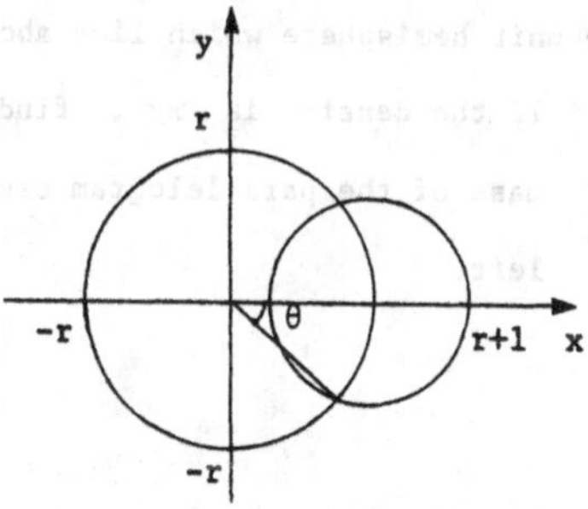

Center the larger circle at the origin, and center the smaller one at $(r,0)$, then the equations of the circles are $x^2 + y^2 = r^2$ and $(x - r)^2 + y^2 = 1$. Subtracting one equation from the other gives us the point of intersection: $x = (2r^2 - 1)/2r$. Therefore, $\cos\theta = (2r^2 - 1)/2r^2$, and by the formula derived in part (a), the surface area is $2\pi r^2(1 - (2r^2 - 1)/2r^2) = 2\pi r^2(1/2r^2) = \pi$. This is surprising because no matter how large r is, the surface area cut out is always π , independent of r .

33. As explained in the solution to Example 5, $z = \pm\sqrt{[f(x)]^2 - y^2}$ and the region of integration is given by $a \leqslant x \leqslant b$ and $-f(x) \leqslant y \leqslant f(x)$. The double integral of z over D gives the volume under the graph of z and since half of z lies below the xy-plane, the volume is $2\int_a^b\int_{-f(x)}^{f(x)} \sqrt{[f(x)]^2 - y^2} \, dy \, dx$. By formula 38 of the integration table, we get $2\int_a^b\{[(y/2)\sqrt{[f(x)]^2 - y^2} + ((f(x))^2/2) \sin^{-1}(y/f(x))] |_{y=-f(x)}^{f(x)}\}dx = 2\int_a^b((f(x))^2/2)(\pi/2 - (-\pi/2)) \, dx = \pi\int_a^b[f(x)]^2dx$. This is exactly the same formula that was derived using one-variable calculus.

SECTION QUIZ

1. Find the volume of the solid bounded by $z = xy + 3$, the xy-plane, and the cylinder $(x - 2)^2 + (y - 2)^2 = 1$.

2. Find the average value of the unit hemisphere which lies above the xy-plane

3.   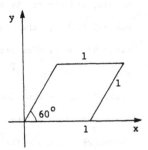   If the density is xy , find the center of mass of the parallelogram sketched at the left.

4. Find the surface area of $z = x + y + 4$ over the region $y \geqslant x^2$ and $y \leqslant 9$.

5. Herbie, the health nut, refuses to drink anything except melted snow water. Even when the weather is warm, Herbie imports snow from the North Pole. Last night, a blizzard left a fresh blanket of snow on the ground with height $2 + (x \sin y)/100$ in Herbie's backyard: $[0,100] \times [0,50]$.

    (a) If one cubic unit represents one gallon of water, how many gallon containers are needed to collect the water for Health Nut Herbie?

    (b) Herbie decides to cover the snow and collect it later. What size tarp does he need? Express your answer as a double integral.

ANSWERS TO PREREQUISITE QUIZ

1. 1/3

2. $\bar{x} = \int_a^b x\, f(x)dx / \int_a^b f(x)dx$; $\bar{y} = (1/2)\int_a^b [f(x)]^2 dx / \int_a^b f(x)dx$

3. $2\pi \int_a^b f(x)\sqrt{1 + [f'(x)]^2}\, dx$

4. $2\pi \int_a^b x\sqrt{1 + [f'(x)]^2}\, dx$

5. 3/2

ANSWERS TO SECTION QUIZ

1. 11π

2. 2/3

3. $(19/20, 7\sqrt{3}/20)$

4. $12\sqrt{3}$

5. (a) $10050 - 50 \cos 50$

    (b) $(1/100)\int_0^{100}\int_0^{50} [10000 + \sin^2 y + x^2\cos^2 y]^{1/2} dy\, dx$

17.4 Triple Integrals

PREREQUISITES

1.   Recall how to compute double integrals (Section 17.2).

PREREQUISITE QUIZ

1.   Compute the following integrals:

(a) $\int_0^2 \int_0^{x^2} xy \, dy \, dx$

(b) $\int_1^2 \int_{-y}^1 y \, dx \, dy$

2.

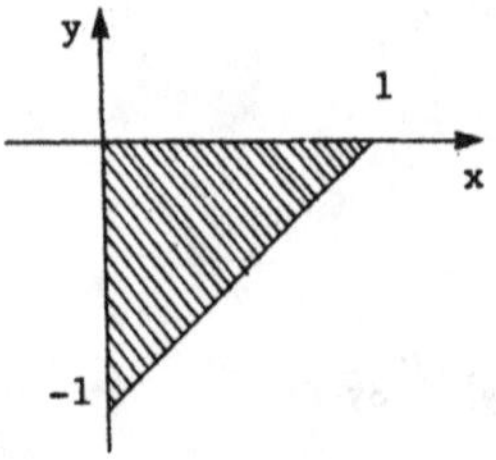

Compute $\iint_D (x + y) dy \, dx$ for the region D shown at the left.

GOALS

1.   Be able to evaluate a triple integral by using iterated integrals.

2.   Be able to change the order of integration in triple integrals.

STUDY HINTS

1.   Triple integrals. As with all of the integration theory presented up to this point, the triple integral is that unique number that lies between all lower sums and all upper sums.

2.   Iterated integral. As with double integrals, the triple integral may be reduced to iterated integrals. Again, the variables may be integrated in any order.

3. __Region types__. W is type I if it is bounded by two surfaces which are functions of x and y . The region D over which W falls in the xy-plane may be either type 1 or 2. Type II regions interchange the roles of x and z . Type III regions interchange the roles of y and z . As with double integrals, knowing the region type aids in choosing the limits of integration; however, knowing the name of the region type does not help in doing computations.

4. __Balls__. Example 3 demonstrates that the unit ball is a type I region. Written as $-1 \leqslant z \leqslant 1, -\sqrt{1 - z^2} \leqslant y \leqslant \sqrt{1 - z^2}$, and $-\sqrt{1 - z^2 - y^2} \leqslant x \leqslant \sqrt{1 - z^2 - y^2}$, the ball is a type II region. Similarly, the type III ball is described by $-1 \leqslant x \leqslant 1$, $-\sqrt{1 - x^2} \leqslant z \leqslant \sqrt{1 - x^2}$, $-\sqrt{1 - x^2 - z^2} \leqslant y \leqslant \sqrt{1 - x^2 - z^2}$. The solution of Example 3 uses a type 1 description for D . Using a type 2 description would generate three more descriptions of W , the unit ball.

SOLUTIONS TO EVERY OTHER ODD EXERCISE

1. The first method we use is to integrate with respect to x , then y , then z . The second method is integrating with respect to y , then z , then x . Any other order may be used and the answer is always the same.

$$\iiint_W (2x + 3y + z)dx\, dy\, dz = \int_0^1\int_{-1}^1\int_1^2 (2x + 3y + z)dx\, dy\, dz =$$
$$\int_0^1\int_{-1}^1 (x^2 + 3xy + xz)\big|_{x=1}^2 \, dy\, dz = \int_0^1\int_{-1}^1 (3 + 3y + z)dy\, dz =$$
$$\int_0^1 (3y + 3y^2/2 + yz)\big|_{y=-1}^1 \, dz = \int_0^1 (6 + 2z)dz = (6z + z^2)\big|_0^1 = 7 \ .$$

The second method gives $\int_1^2\int_0^1\int_{-1}^1 (2x + 3y + z) \, dy\, dz\, dx =$
$$\int_1^2\int_0^1 (2xy + 3y^2/2 + yz)\big|_{y=-1}^1 \, dz\, dx = \int_1^2\int_0^1 (4x + 2z)dz\, dx =$$
$$\int_1^2 (4xz + z^2)\big|_{z=0}^1 \, dx = \int_1^2 (4x + 1)dx = (2x^2 + x)\big|_1^2 = 7 \ .$$

5.

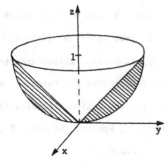

W is type I.

The region is sketched at the left. It lies over the region D , which is a circle of radius 1 centered at the origin. Thus, the region W can be described by $-1 \leqslant x \leqslant 1$, $-\sqrt{1 - x^2} \leqslant y \leqslant \sqrt{1 - x^2}$, and $x^2 + y^2 \leqslant z \leqslant \sqrt{x^2 + y^2}$. As described,

9.

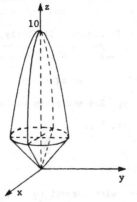

The volume is $\iiint_W dx\, dy\, dz$. The projection of the intersection of the two surfaces is $x^2 + y^2 = 10 - x^2 - 2y^2$, i.e., $x^2/(\sqrt{5})^2 + y^2/(\sqrt{10/3})^2 = 1$. Thus, the volume equals

$$\int_{-\sqrt{5}}^{\sqrt{5}} \int_{-\sqrt{(10-2x^2)/3}}^{\sqrt{(10-2x^2)/3}} \int_{x^2+y^2}^{10-x^2-2y^2} dz\, dy\, dx =$$

$$\int_{-\sqrt{5}}^{\sqrt{5}} \int_{-\sqrt{(10-2x^2)/3}}^{\sqrt{(10-2x^2)/3}} (10 - 2x^2 - 3y^2)\, dy\, dx =$$

$$= \int_{-\sqrt{5}}^{\sqrt{5}} (10y - 2x^2 y - y^3) \left.\right|_{y=-\sqrt{(10-2x^2)/3}}^{\sqrt{(10-2x^2)/3}} dx = 4\int_{-\sqrt{5}}^{\sqrt{5}} ((10 - 2x^2)/3)^{3/2}\, dx =$$

$$4(2/3)^{3/2} \int_{-\sqrt{5}}^{\sqrt{5}} (5 - x^2)^{3/2}\, dx = 4(2/3)^{3/2} [(x/4)(5 - x^2)^{3/2} +$$

$$(15/4)((x/2)\sqrt{5 - x^2} + (5/2) \sin^{-1} (x/\sqrt{5}))] \left.\right|_{-\sqrt{5}}^{\sqrt{5}} = 25\sqrt{2/3}\,\pi .$$

13. $\int_0^1\int_1^2\int_2^3 \cos[\pi(x + y + z)]\,dx\,dy\,dz = \int_0^1\int_1^2 [\sin(\pi(x + y + z))/\pi]\,\big|_{x=2}^3\,dy\,dz =$

$(1/\pi)\int_0^1\int_1^2 [\sin(\pi(3 + y + z)) - \sin(\pi(2 + y + z))]\,dy\,dz =$

$(1/\pi)\int_0^1 [-\cos(\pi(3 + y + z))/\pi + \cos(\pi(2 + y + z))/\pi]\,\big|_{y=1}^2\,dz =$

$(1/\pi^2)\int_0^1 [-\cos(\pi(5 + z)) + 2\cos(\pi(4 + z)) - \cos(\pi(3 + z))]\,dz =$

$(1/\pi^2)[-\sin(\pi(5 + z))/\pi + 2\sin(\pi(4 + z))/\pi - \sin(\pi(3 + z))/\pi]\,\big|_0^1 = 0$.

17. $\int_0^\pi\int_0^1\int_0^{1-y} x^2 \cos z\,dx\,dy\,dz = \int_0^\pi\int_0^1 [(x^3 \cos z)/3\,\big|_{x=0}^{1-y}]\,dy\,dz =$

$(1/3)\int_0^\pi\int_0^1 (1 - y)^3 \cos z\,dy\,dz$. Let $u = 1 - y$ to get

$(-1/3)\int_0^\pi\int_{-1}^0 u^3 \cos z\,du\,dz = (-1/3)\int_0^\pi [(u^4 \cos z)/4\,\big|_{u=-1}^0]\,dz =$

$(1/12)\int_0^\pi \cos z\,dz = (1/12)(-\sin z)\,\big|_0^\pi = 0$. (This problem could have

been done in one step by integrating in z first.)

21. Let the box W be $[a,b] \times [c,d] \times [p,q]$, then

$\int_a^b\int_c^d\int_p^q F(x,y)\,dz\,dy\,dx$. When integrating in z , $F(x,y)$ is held

constant, so the integral is $\int_a^b\int_c^d [zF(x,y)\,|_{z=p}^q]\,dy\,dx =$

$(q - p)\int_a^b\int_c^d F(x,y)\,dy\,dz$. Thus, the triple integral of f over a

box W is the double integral of F over the base of the box, times

the height of the box.

25. Let W be the solid whose volume we wish to compute. Then the volume

is $\iiint_W dx\,dy\,dz = \iiint_W dz\,dx\,dy$ by changing the order of integration.

Using either formula (3) or (4), the integral becomes $\iint_D\int_0^{f(x,y)} dz\,dx\,dy =$

$\iint_D (z\,|_{z=0}^{f(x,y)})\,dx\,dy = \iint_D f(x,y)\,dx\,dy$, which is the double integral of

f over D .

SECTION QUIZ

1. Evaluate $\iiint_W(xz + yz)\,dx\,dy\,dz$, where W is the box $[0,1] \times$

$[1,3] \times [-1,1]$.

2.   (a)   Evaluate $\int_1^2 \int_{-2}^1 \int_0^3 \, dy \, dz \, dx$.

(b)   Sketch the region of integration.

(c)   Give a geometric interpretation of the integral in part (a).

3.   Define the triple integral in terms of upper and lower sums.

4.   Express the volume between $z = xy$ and the region $[-3,2] \times [-1,1]$

on the xy-plane as a sum of triple integrals.

5.   Suppose a college student's state of alertness during a lecture is a

function of   s ,   how much sleep he or she had the night before, in

hours;   i ,   the interest level of the professor's lecture; and   g ,   the

student's grade point average.   A student's Falling Asleep Index for

Lectures (FAIL) is determined by $1 - [\int_0^a \int_0^b \int_0^c \, s^2 ig \, dg \, di \, ds] \,/$

$[\int_0^{10} \int_0^{10} \int_0^4 \, s^2 ig \, dg \, di \, ds]$.

(a)   What is a student's FAIL if $a = 5$, $b = 2$,   and $c = 3.5$?

(b)   If $b = 2$ and $c = 3.5$ throughout the semester, how many hours

of sleep is needed to keep FAIL above   0.5 ?   (A FAIL of   0.5

indicates a   50%   probability of falling asleep.   A greater FAIL

indicates a greater likelihood of falling asleep.)

ANSWERS TO PREREQUISITE QUIZ

1.   (a)   16/3

(b)   23/6

2.   0

ANSWERS TO SECTION QUIZ

1.   0

2.   (a)   9

     (b)

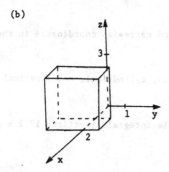

     (c)   It is the volume of the box sketched in part (b).

3.   The precise number between all lower sums and all upper sums.

4. $\int_0^2\int_0^1\int_0^{xy} dz\ dy\ dx + \int_{-3}^0\int_{-1}^0\int_0^{xy} dz\ dy\ dx + \int_{-3}^0\int_0^1\int_{xy}^0 dz\ dy\ dx + \int_0^2\int_{-1}^0\int_{xy}^0 dz\ dy\ dx$

5.   (a)   12751/12800

     (b)   200 $\sqrt[3]{10/49}$ (Impossible! The student will probably fall asleep.)

17.5 Integrals in Polar, Cylindrical, and Spherical Coordinates

PREREQUISITES

1. Recall how to convert between polar and cartesian coordinates in the
    plane (Section 5.1).

2. Recall how to convert between cartesian, cylindrical, and spherical
    coordinates in space (Section 14.5).

3. Recall how to compute double and triple integrals (Sections 17.2 and 17.4).

PREREQUISITE QUIZ

1. Let $P = (2,1,-1)$ be the cartesian coordinates of a point in space.
    (a) What are the cylindrical coordinates of P ?
    (b) What are the spherical coordinates of P ?

2. Evaluate the triple integral $\int_0^1 \int_0^{\pi/2} \int_0^{\pi/4} \rho^2 \sin\theta \cos(\phi/2) \, d\theta \, d\phi \, d\rho$.

3. Evaluate the double integral $\int_0^{\pi/2} \int_0^1 r^2 \cos\theta \, dr \, d\theta$.

GOALS

1. Be able to integrate multiple integrals in polar coordinates, cylindrical
    coordinates, and spherical coordinates.

2. Be able to evaluate the Gaussian integral.

STUDY HINTS

1. <u>Polar coordinates</u>. Recall that dx dy converts to r dr dθ (if you
    remember that arc length is r dθ and draw Fig. 17.5.1, you can always
    remember this), so if we describe D in terms of r and θ to get
    D′ and substitute, the resulting expression is $\iint_D f(x,y) dx \, dy =$
    $\iint_{D'} f(r \cos\theta, r \sin\theta) r \, dr \, d\theta$.

2.   **Gaussian integral.** You should memorize the fact that $\int_{-\infty}^{\infty} \exp(-x^2)dx =$ $\sqrt{\pi}$. In addition, you should be able to manipulate this equation to compute similar integrals. See Example 2.

3.   **Cylindrical coordinates.** As with polar coordinates, dx dy changes to r dr dθ , so if we describe W in terms of r , θ , and z to get W' and substitute, the resulting expression is $\iiint_W f(x,y,z)dx\,dy\,dz =$ $\iiint_W \, 'f(r\cos\theta, r\sin\theta, z)r\,dr\,d\theta\,dz$.

4.   **Spherical coordinates.** You should memorize the fact that dx dy dz converts to $\rho^2\sin\phi\,d\rho\,d\theta\,d\phi$ (the volume of the box in Fig. 17.5.4). By expressing W in terms of ρ , θ , and φ to get W' , substitution gives $\iiint_W f(x,y,z)dx\,dy\,dz = \iiint_W \, 'f(\rho\sin\phi\cos\theta, \rho\sin\phi\sin\theta,$ $\rho\cos\phi)\rho^2\sin\phi\,d\rho\,d\theta\,d\phi$. When you set up the limits of integration, remember that φ is measured from the "north pole," <u>not</u> from the "equator." Also, recall that θ is measured in a counterclockwise direction.

5.   **Choosing an integration method.** If $x^2 + y^2$ occurs in the integrand of a double integral, it is generally a good idea to use polar coordinates. Cylindrical coordinates are generally used if $x^2 + y^2$ appears in the integrand of a triple integral. And finally, if $x^2 + y^2 + z^2$ appears in a triple integral, it is wise to try spherical coordinates.

SOLUTIONS TO EVERY OTHER ODD EXERCISE

1.   The region D can be described by $0 \leqslant r \leqslant 2$, $0 \leqslant \theta \leqslant 2\pi$; thus $\iint_D (x^2 + y^2)^{3/2}dx\,dy = \int_0^{2\pi}\int_0^2 r(r^2)^{3/2}dr\,d\theta = \int_0^{2\pi}\int_0^2 r^4\,dr\,d\theta =$ $\int_0^{2\pi}(r^5/5)|_{r=0}^{2}\,d\theta = \int_0^{2\pi}(2^5/5)d\theta = 64\pi/5$.

5.   The disk is described by $0 \leqslant r \leqslant 4$ and $0 \leqslant \theta \leqslant 2\pi$, so $\iint_D(x^2 + y^2)dx\,dy = \int_0^{2\pi}\int_0^4(r^2)r\,dr\,d\theta = \int_0^{2\pi}(r^4/4)|_{r=0}^{1}\,d\theta = (1/4)\int_0^{2\pi}d\theta = \pi/2$.

9.   can be described in spherical coordinates by $0 \leq \rho \leq 1$,

$0 \leq \phi \leq \pi$, $0 \leq \theta \leq 2\pi$; thus $\iiint_W (dx \; dy \; dz / \sqrt{1 + x^2 + y^2 + z^2})$ =

$\int_0^1 \int_0^\pi \int_0^{2\pi} (\rho^2 \sin \phi \; d\theta \; d\phi \; d\rho / \sqrt{1 + \rho^2}) = \int_0^1 [(2\pi\rho^2 / \sqrt{1 + \rho^2}) \int_0^\pi \sin \phi \; d\phi] d\rho$ =

$\int_0^1 (4\pi \; \rho^2 / \sqrt{1 + \rho^2}) d\rho = 4\pi \int_0^1 [(\rho^2 + 1 - 1) / \sqrt{1 + \rho^2}] \; d\rho = 4\pi [\int_0^1 \sqrt{1 + \rho^2} \; d\rho$ -

$\int_0^1 (d\rho / \sqrt{1 + \rho^2})] = 4\pi [(\rho/2)\sqrt{1 + \rho^2} + (1/2) \ln(\rho + \sqrt{1 + \rho^2})$ -

$\ln(\rho + \sqrt{1 + \rho^2})] \Big|_0^1 = 2\pi [\sqrt{2} - \ln(1 + \sqrt{2})]$.

13.

The region is described by

$-\sqrt{1 - x^2 - y^2} \leq z \leq \sqrt{1 - x^2 - y^2}$,

$-\sqrt{1/4 - x^2} \leq y \leq \sqrt{1/4 - x^2}$, and

$-1/2 \leq x \leq 1/2$; thus, the volume is

$\int_{-1/2}^{1/2} \int_{-\sqrt{1/4-x^2}}^{\sqrt{1/4-x^2}} \int_{-\sqrt{1-x^2-y^2}}^{\sqrt{1-x^2-y^2}} dz \; dy \; dx$ =

$\int_{-1/2}^{1/2} \int_{-\sqrt{1/4-x^2}}^{\sqrt{1/4-x^2}} (2\sqrt{1 - x^2 - y^2}) \; dy \; dx$. The region $-\sqrt{1/4 - x^2} \leq y \leq$

$\sqrt{1/4 - x^2}$ and $-1/2 \leq x \leq 1/2$ can be described in polar coordinates

by $0 \leq r \leq 1/2$ and $0 \leq \theta \leq 2\pi$. Therefore, we have volume =

$\int_0^{2\pi} \int_0^{1/2} 2\sqrt{1 - r^2} \; r \; dr \; d\theta = \int_0^{2\pi} [-(2/3)(1 - r^2)^{3/2}] \Big|_0^{1/2} \; d\theta$ =

$(\pi/6)(8 - 3\sqrt{3})$.

17. We know that $\int_{-\infty}^\infty c \; \exp(-x^2/\sigma) dx = c \int_{-\infty}^\infty \exp(-x^2/\sigma) dx$. Let $y = x/\sqrt{\sigma}$,

so the integral becomes $\lim_{a \to \infty} c \int_{-a}^a \exp(-x^2/\sigma) dx = \lim_{a \to \infty} c \int_{-a/\sqrt{\sigma}}^{a/\sqrt{\sigma}} \exp(-y^2) dy \sqrt{\sigma}$ =

$\lim_{a \to \infty} c\sqrt{\sigma} \int_{-a/\sqrt{\sigma}}^{a/\sqrt{\sigma}} \exp(-y^2) dy$. By the Gaussian integral, this is

$c\sqrt{\sigma} \int_{-\infty}^\infty \exp(-y^2) dy = c\sqrt{\pi\sigma}$. Therefore, the normalizing constant is

$c = 1/\sqrt{\pi\sigma}$.

21. 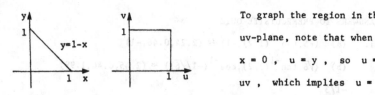 To graph the region in the
uv-plane, note that when
$x = 0$, $u = y$, so $u =$
uv , which implies u = 0

or $v = 1$. So when $v = 1$, $u = y$ is in the interval $0 \leqslant u \leqslant 1$.
When $y = 0$, $u = x$ and $0 = uv$, which implies $u = 0$ or $v = 0$
for $0 \leqslant u \leqslant 1$. When $y = 1 - x$, $u = x + (1 - x) = 1$ and $y = v$
for $0 \leqslant v \leqslant 1$. Thus, the region of integration is $0 \leqslant u \leqslant 1$ and
$0 \leqslant v \leqslant 1$.

The integrand is $\exp[y/(x + y)] = \exp[uv/u] = e^v$. Rearrangement
gives $x = u - uv$ and $y = uv$, so $\| |\partial(x,y)/\partial(u,v)| \| = $
$\begin{vmatrix} 1 - v & -u \\ v & u \end{vmatrix} = u$. Therefore, the integral is $\int_0^1\int_0^1 e^v(u)\, du\, dv = $
$\int_0^1 [(u^2/2)e^v|_{u=0}^1]\, dv = (1/2)e^v|_0^1 = (e - 1)/2$.

SECTION QUIZ

1. Discuss the evaluation of $\int_{-\infty}^{\infty} \exp(3x^2)\, dx$.

2. Evaluate the triple integral $\iiint_W ((x^2 + y^2)/z)dx\, dy\, dz$, where W is
the cylinder $x^2 + y^2 \leqslant 9$ and $1 \leqslant z \leqslant 2$.

3. Evaluate the triple integral $\iiint_W ((x^2 + y^2)/(x^2 + y^2 + z^2))dz\, dx\, dy$,
where W is the sphere $x^2 + y^2 + z^2 \leqslant 4$.

4. Evaluate the double integral $\iint_D x\sqrt{x^2 + y^2}\, dy\, dx$, where D is the disk
$x^2 + y^2 \leqslant 1$.

5. The big bad wolf lived in the center of the forest. Since the big bad
wolf's big, sharp teeth were used to eat rabbits, the density of the rabbit
population was $x^4 + 2x^2y^2 + y^4$ at (x,y) . Compute the rabbit population
within radius R of the big bad wolf's home, located at $(0,0)$.

ANSWERS TO PREREQUISITE QUIZ

1.   (a) $(\sqrt{5},\tan^{-1}(1/2),-1) \approx (2.24,0.46,-1)$

     (b) $(\sqrt{6},\tan^{-1}(1/2),\cos^{-1}(-1/\sqrt{6})) \approx (2.45,0.46,1.99)$

2. $-1/3$

3. $1/3$

ANSWERS TO SECTION QUIZ

1.   The integral is infinite.   The Gaussian integral requires a negative
     exponent.

2. $(81\pi/2) \ln 2$

3. $64\pi/3$

4. 0

5. $\pi R^6/6$

17.6 Applications of Triple Integrals

PREREQUISITES

1. Recall how to compute an average or center of mass using double
    integration (Section 17.3).

2. Recall how to compute triple integrals in cartesian coordinates
    (Section 17.4).

3. Recall how to compute triple integrals in spherical coordinates
    (Section 17.5).

PREREQUISITE QUIZ

1. Find the average of $f(x,y) = xy$ over the rectangle $[0,1] \times [0,2]$.

2. Compute the center of mass for the plate with density $\rho(x,y) = xy$
    if the plate can be described by $0 \leqslant x \leqslant 1$ and $0 \leqslant y \leqslant 2$.

3. Evaluate $\int_0^1 \int_0^1 \int_{-1}^1 z(x^2 + y^2) dx\, dy\, dz$.

4. Evaluate $\iiint_W z\, dx\, dy\, dz$, where W is the unit sphere.

GOALS

1. Be able to apply triple integrals for computing volumes, centers of
    mass, and averages.

STUDY HINTS

1. Volume. If W is the region under the graph of $z = f(x,y)$, then
    $\iiint_W dx\, dy\, dz = \iint_D z\, dx\, dy$. Since $z = f(x,y)$, the original integral
    becomes $\iint_D f(x,y) dx\, dy$, which is the formula given in the first half
    of this chapter.

2. <u>Mass</u>. $\iiint_W \rho(x,y,z)\ dx\ dy\ dz$ is just an extension of the one- and two-
    variable integrals.

3. <u>Center of mass</u>. The formulas are just an extension of the one- and
    two-variable cases.

4. <u>Averages</u>. With the addition of an extra variable, we now divide the
    integral by its volume, rather than the area for the two-variable case.

5. <u>Gravitational potential</u>. The chapter ends with a discussion of the
    application of triple integrals to gravity. Although you don't need
    to understand the physical theory behind the discussion, the mathematics
    should be clear to you. You probably will not need to reproduce the
    discussion for an exam.

SOLUTIONS TO EVERY OTHER ODD EXERCISE

1. The mass is given by $\iiint_W \rho(x,y,z)\,dx\ dy\ dz$.

    (a) Let ρ be the constant mass density. Then, the mass is

    $\rho\int_0^{1/2}\int_0^1\int_0^2 dz\ dy\ dx = \rho(1/2)(1)(2) = \rho$.

    (b) The mass is $\int_0^{1/2}\int_0^1\int_0^2 (x^2 + 3y^2 + z^2 + 1)dz\ dy\ dx =$

    $\int_0^{1/2}\int_0^1 (zx^2 + 3y^2z + z^3/3 + z)\big|_{z=0}^2\ dy\ dx = \int_0^{1/2}\int_0^1 (2x^2 + 6y^2 +$

    $14/3)dy\ dx = \int_0^{1/2} (2yx^2 + 2y^3 + 14y/3)\big|_{y=0}^1\ dx = \int_0^{1/2} (2x^2 + 20/3)dx =$

    $(2x^3/3 + 20x/3)\big|_0^{1/2} = 1/12 + 10/3 = 41/3$.

5. Let $I = \iiint_W z\ dx\ dy\ dz$. Then, by considering the hemisphere as a type
    I region, we integrate with respect to x last, so $I =$

    $\int_{-1}^1 \int_{-\sqrt{1-x^2}}^{\sqrt{1-x^2}} \int_0^{\sqrt{1-x^2-y^2}} z\ dz\ dy\ dx = (1/2)\int_{-1}^1 \int_{-\sqrt{1-x^2}}^{\sqrt{1-x^2}} (1 - x^2 - y^2)dy\ dx =$

    $(1/2)\int_{-1}^1 (4/3)(1 - x^2)^{3/2}\ dx$ (formula 38 of the integration table) $=$

    $(2/3)[(x/4)(1 - x^2)^{3/2} + (3/4)(x\sqrt{1 - x^2} + (1/2)\ \sin^{-1}x]\big|_{-1}^1$ (formula

    39) $= \pi/4$.

9. The average value of $f(x,y,z)$ is $\iiint_W f(x,y,z)dx\, dy\, dz / \iiint_W dx\, dy\, dz$.

The numerator is $\int_0^2\int_0^4\int_0^6 \sin^2 \pi z \cos^2 \pi x\, dz\, dy\, dx = \int_0^2\int_0^4\int_0^6 ((1 - \cos 2\pi z)/2) \times$

$((1 + \cos 2\pi x)/2)dz\, dy\, dx = (1/4)\int_0^2 (1 + \cos 2\pi x)dx \cdot \int_0^4 dy \cdot \int_0^6 (1 - \cos 2\pi z)dz =$

$(1/4)(x + \sin 2\pi x/2\pi)\big|_0^4 \cdot 4 \cdot (z - \sin 2\pi z/2\pi)\big|_0^6 = (1/4)(4)(4)(6) = 24$. The

denominator is $\int_0^2\int_0^4\int_0^6 dz\, dy\, dx = (2)(4)(6) = 48$. Thus, the average value

is $1/2$.

13. Using the method of Example 6, we have $V = GM/R = (6.67 \times 10^{-11}) \times$

$(3 \times 10^{26})/(2 \times 10^8) = 1.00 \times 10^8 (m/s)^2$.

17. We will place (x_1,y_1,z_1) at $(0,0,R)$ and the spherically symmetric

body will be centered at the origin. According to the text, the gravi-

tational potential due to the region between the concentric spheres

$\rho = \rho_1$ and $\rho = \rho_2$ is GM/R if (x_1,y_1,z_1) lies outside the attrac-

ting body. In this problem, the density is a function of the radius, so

we will integrate over the regions between the spheres $\rho = \rho_0$ and

$\rho = \rho_0 + d\rho$. For each region, the density is constant, and the gravi-

tational potential for each infinitesimal region is Gm_i/R where m_i

is the mass of the ith region. Since the total mass M is Σm_i , the

gravitational potential for the entire body is GM/R .

SECTION QUIZ

1. A region in space is bounded by $z = xy$, $z = x - y$, $y = 3$, $y = -x$,

$x = 0$, and $x = 3$.

(a) Compute the volume of the region.

(b) Find the region's center of mass, assuming constant density.

2.   A region is bounded by the planes $z = x + y$, $y = x + 1$, $x = 2$,
     $z = 0$, $y = 0$, and $x = 0$.

     (a) What is the average value of $x + y + z$ over this region?

     (b) If the density is xyz , what is the mass of the region?

3.   To join the PTA (Party Troupers of America), one must demonstrate an
     aptitude for having a good time. Applicants are given three 100-point
     exams in the categories Eating, Drinking, and Merriment. The point
     scores (E , D , and M , respectively) are combined by the formula
     $S = \int_0^E \int_0^D \int_0^M (edm)^{-5/6} dm \, dd \, de$.

     (a) What is the combined perfect score, i.e., what is S when $E = D =$
          $M = 100$?

     (b) One needs 75% of the score in part (a) to qualify for admission
          into the PTA. Will 60 points in each category meet this requiremen

ANSWERS TO PREREQUISITE QUIZ

1.   1/2

2.   (2/3, 4/3)

3.   2/3

4. $\pi/4$

ANSWERS TO SECTION QUIZ

1.   (a)   333/8

     (b)   (327/185, 147/37, 57/37)

2.   (a)   37/9

     (b)   5549/180

3.   (a)   2160

     (b)   Yes; it is over 77% .

17.R Review Exercises for Chapter 17

SOLUTIONS TO EVERY OTHER ODD EXERCISE

1. $\int_2^3 \int_4^8 [x^3 + \sin(x + y)]\,dx\,dy = \int_2^3 \{ [x^4/4 - \cos(x + y)] \mid_{x=4}^8 \}\,dy =$

   $\int_2^3 [960 - \cos(8 + y) + \cos(4 + y)]\,dy = [960y - \sin(8 + y) +$

   $\sin(4 + y)] \mid_2^3 = 960 - \sin(11) + \sin(10) + \sin(7) - \sin(6) \approx 961.4$.

5.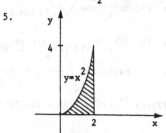

   The region may be described as the following
   type 1 region: $0 \leqslant y \leqslant x^2$ and $0 \leqslant x \leqslant 2$.

   Thus, $\iint_D (x^3 + y^2 x)\,dx\,dy = \int_0^2 \int_0^{x^2} (x^3 +$

   $y^2 x)\,dy\,dx = \int_0^2 [(x^3 y + y^3 x/3) \mid_{y=0}^{x^2}]\,dx =$

   $\int_0^2 (x^5 + x^7/3)\,dx = (x^6/6 + x^8/24) \mid_0^2 =$

   $2^6/(2)(3) + 2^8/(2^3)(3) = 32/3 + 32/3 = 64/3$.

9. $\int_0^1 \int_0^x \int_0^y xyz\,dz\,dy\,dx = \int_0^1 \int_0^x (xyz^2/2) \mid_{z=0}^y\,dy\,dx = \int_0^1 \int_0^x (xy^3/2)\,dy\,dx =$

   $\int_0^1 (xy^4/8) \mid_{y=0}^x\,dz = \int_0^1 (x^5/8)\,dx = (x^6/48) \mid_0^1 = 1/48$.

13. The volume of the solid is $\iiint_W dx\,dy\,dz$.

   $\int_{-a}^a \int_{-b\sqrt{1-x^2/a^2}}^{b\sqrt{1-x^2/a^2}} \int_{-c\sqrt{1-x^2/a^2-y^2/b^2}}^{c\sqrt{1-x^2/a^2-y^2/b^2}} dz\,dy\,dx =$

   $2c \int_{-a}^a \int_{-b\sqrt{1-x^2/a^2}}^{b\sqrt{1-x^2/a^2}} \sqrt{1 - x^2/a^2 - y^2/b^2}\,dy\,dx$. Factor out $1/b$ to get

   $2c \int_{-a}^a \int_{-b\sqrt{1-x^2/a^2}}^{b\sqrt{1-x^2/a^2}} (1/b)\sqrt{b^2 - b^2 x^2/a^2 - y^2}\,dy\,dx$. Use formula 38 of the

   integration table to get $2c \int_{-a}^a \Big\{ (1/b) [(y/2)\sqrt{b^2 - b^2 x^2/a^2 - y^2} +$

   $((b^2 - b^2 x^2/a^2)/2)\sin^{-1}(y/\sqrt{b^2 - b^2 x^2/a^2})] \mid_{y=-b\sqrt{1-x^2/a^2}}^{b\sqrt{1-x^2/a^2}} \Big\}\,dx =$

   $2c \int_{-a}^a (1/b)((b^2 - b^2 x^2/a^2)/2)\pi dx = \pi cb \int_{-a}^a (1 - x^2/a^2)\,dx = \pi bc(x -$

   $x^3/3a^2) \mid_{-a}^a = \pi bc(4a/3) = (4/3)\pi abc$.

17.

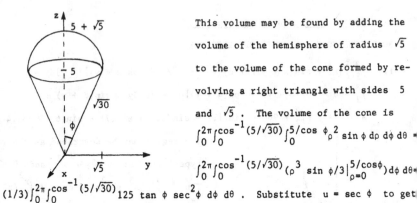

This volume may be found by adding the volume of the hemisphere of radius $\sqrt{5}$ to the volume of the cone formed by revolving a right triangle with sides 5 and $\sqrt{5}$. The volume of the cone is

$$\int_0^{2\pi}\int_0^{\cos^{-1}(5/\sqrt{30})}\int_0^{5/\cos\phi}\rho^2\sin\phi\,d\rho\,d\phi\,d\theta =$$

$$\int_0^{2\pi}\int_0^{\cos^{-1}(5/\sqrt{30})}(\rho^3\sin\phi/3\Big|_{\rho=0}^{5/\cos\phi})d\phi\,d\theta =$$

$(1/3)\int_0^{2\pi}\int_0^{\cos^{-1}(5/\sqrt{30})}125\tan\phi\sec^2\phi\,d\phi\,d\theta$. Substitute $u = \sec\phi$ to get

$(125/3)\int_0^{2\pi}[(\sec^2\phi/2)\Big|_{\phi=0}^{\cos^{-1}(5/\sqrt{30})}]\,d\theta = (125/6)\int_0^{2\pi}(6/5-1)d\theta = 2\pi(25/6) =$

$25\pi/3$.

The calculations simplify if the sphere $x^2 + y^2 + (z-5)^2 = 5$ is centered at the origin. Then the volume of the hemisphere is

$$\int_0^{2\pi}\int_{\pi/2}^{\pi}\int_0^{\sqrt{5}}\rho^2\sin\phi\,d\rho\,d\phi\,d\theta = \int_0^{2\pi}\int_{\pi/2}^{\pi}[\rho^3\sin\phi/3\Big|_{\rho=0}^{\sqrt{5}}]\,d\phi\,d\theta =$$

$(5\sqrt{5}/3)\int_0^{2\pi}\int_{\pi/2}^{\pi}\sin\phi\,d\phi\,d\theta = (5\sqrt{5}/3)\int_0^{2\pi}[(-\cos\phi)\Big|_{\phi=\pi/2}^{\pi}]\,d\theta =$

$(5\sqrt{5}/3)\int_0^{2\pi}d\theta = 10\sqrt{5}\pi/3$. Thus, the total volume is $(25 + 10\sqrt{5})\pi/3$.

Alternatively, one may find it easier to use cylindrical coordinates to compute the volume of the cone or simply use $V = \pi r^2 h/3$. Similarly, the volume of the sphere may be computed with $V = (1/2)(4\pi r^3/3)$.

21.

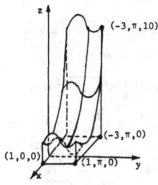

The volume under the graph of $f(x,y)$

is $\int_a^b \int_c^d f(x,y)\ dy\ dx$. $\int_{-3}^1 \int_0^\pi [x^2 +$

$\sin(2y) + 1]\ dy\ dx = \int_{-3}^1 \{[y(x^2 + 1) -$

$(1/2)\cos(2y)]\ |_{y=0}^\pi \}\ dx =$

$\int_{-3}^1 \pi(x^2 + 1)dx = \pi(x^3/3 + x)|_{-3}^1 =$

$40\pi/3$.

25.

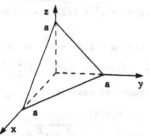

When the cut is made by the plane

$x + y + z = a$, the volume of the

solid below that plane is

$\int_0^a \int_0^{a-x} \int_0^{a-x-y} dz\ dy\ dx =$

$\int_0^a \int_0^{a-x} (a - x - y)dy\ dx =$

$\int_0^a \{[(a - x)y - y^2/2]\ |_{y=0}^{a-x}\}\ dx =$

$\int_0^a [(a - x)^2/2]\ dx$. Substitute $u = a - x$ to get $-(1/2)\int_a^0 u^2 du =$

$-(1/2)(u^3/3)|_{-a}^0 = a^3/6$. Thus, the volume for the entire solid, when

$a = 1$, is $1/6$. If the solid is to be cut into n equal volumes,

then the volume under $x + y + z = a$ should be $k/6n$ where k is an

integer such that $1 \leqslant k \leqslant n - 1$. Therefore, $a^3/6 = k/6n$ implies

that cuts should be made in the planes $x + y + z = \sqrt[3]{k/n}$.

29. In cylindrical coordinates, the "dish" is described by $0 \leqslant r \leqslant 1$,

$0 \leqslant \theta \leqslant 2\pi$, $-\sqrt{25/4 - r^2} + 2 \leqslant z \leqslant 0$. The mass of the "dish" is

$\rho \int_0^1 \int_0^{2\pi} \int_{-\sqrt{25/4-r^2}+2}^0 r\ dz\ d\theta\ dr = \rho \int_0^1 \int_0^{2\pi} (r\sqrt{25/4 - r^2} - 2r)d\theta\ dr =$

$2\pi\rho \int_0^1 (r\sqrt{25/4 - r^2} - 2r)dr$. Let $u = 25/4 - r^2$ to get $2\pi\rho\ [\int_{25/4}^{21/4} \sqrt{u}\ du/$

$(-2) - \int_0^1 2r\ dr] = 2\pi\rho\ [-u^{3/2}/3|_{25/4}^{21/4} - r^2|_0^1] = \pi\rho(101 - 21\sqrt{21})/12$. The

numerator of $\bar{x}$ is $\rho \int_0^1 \int_0^{2\pi} \int_{-\sqrt{25/4-r^2}+2}^0 r^2 \cos\theta\ dz\ d\theta\ dr =$

29. (continued)

$$\rho\int_0^1\int_0^{2\pi}(\sqrt{25/4 - r^2} + 2)r^2 \cos\theta \ d\theta \ dr = \rho\int_0^1[(\sqrt{25/4 - r^2} + 2) \times$$

$r^2 \sin\theta\big|_{\theta=0}^{2\pi}] \ dr = 0$. The numerator of $\bar{y}$ is

$$\rho\int_0^1\int_0^{2\pi}\int_{-\sqrt{25/4-r^2}+2}^0 r^2\sin\theta \ dz \ d\theta \ dr = \rho\int_0^1\int_0^{2\pi}(\sqrt{25/4 - r^2} + 2) \times$$

$r^2 \sin\theta \ d\theta \ dr = -\rho\int_0^1[(\sqrt{25/4 - r^2} + 2)r^2 \cos\theta\big|_{\theta=0}^{2\pi} \ dr = 0$. The

numerator of $\bar{z}$ is $\rho\int_0^1\int_0^{2\pi}\int_{-\sqrt{25/4-r^2}+2}^0 rz \ dz \ d\theta \ dr = \rho\int_0^1\int_0^{2\pi}(rz^2/$

$2\big|_{z=-\sqrt{25/4-r^2}+2}^0)d\theta \ dr = (\rho/2)\int_0^1\int_0^{2\pi} r(-41/4 + r^2 + 4\sqrt{25/4 - r^2})d\theta \ dr =$

$\pi\rho\int_0^1(-41r/4 + r^3 + 4r\sqrt{25/4 - r^2})dr$. Let $u = 25/4 - r^2$ to get

$\pi\rho[(-41r^2/8 + r^4/4)\big|_0^1 - 2\int_{25/4}^{21/4}\sqrt{u} \ du] = \pi\rho[-39/8 - (4u^{3/2}/3)\big|_{25/4}^{21/4}] =$

$\pi\rho(383 - 84\sqrt{21})/24$. Therefore, the center of mass is

$(0,0,(393 - 84\sqrt{21})/(202 - 42\sqrt{21})) \approx (0,0,-0.203)$.

33. The surface area is given by the formula $\iint_D\sqrt{1 + f_x^2 + f_y^2} \ dx \ dy$. In

polar coordinates, D is described by $0 \leqslant r \leqslant 1$, $0 \leqslant \theta \leqslant 2\pi$. $f_x =$

$2x$ and $f_y = -2y$, so the integrand is $\sqrt{1 + 4x^2 + 4y^2} = \sqrt{1 + 4r^2}$.

Therefore, the surface area is $\int_0^{2\pi}\int_0^1 r\sqrt{1 + 4r^2} \ dr \ d\theta$. Let $u =$

$1 + 4r^2$ to get $\int_0^{2\pi}\int_1^5 (\sqrt{u} \ du/8)d\theta = \int_0^{2\pi}(u^{3/2}/12\big|_{u=1}^5)d\theta = \pi(5\sqrt{5} - 1)/6$.

37. In polar coordinates, the region D is described by $1 \leqslant r \leqslant \sqrt{2}$ and

$0 \leqslant \theta \leqslant \pi/2$, so the integral is $\int_0^{\pi/2}\int_1^{\sqrt{2}} (r/r^2)dr \ d\theta =$

$\int_0^{\pi/2}(\ln r\big|_{r=1}^{\sqrt{2}})d\theta = (\pi/2) \ln\sqrt{2} = (\pi/4) \ln(2)$.

41. In each case, the distance from (x,y) to ℓ , the x-axis, is y ;

therefore, $D^2(x,y) = y^2$.

(a) $I = \int_{-1}^1\int_{-2}^2 y^2 \ dy \ dx = \int_{-1}^1 (y^3/3)\big|_{y=-2}^2 \ dx = \int_{-1}^1(16/3)dx =$

$(16x/3)\big|_{-1}^1 = 32/3$.

41. (b) In polar coordinates, we have $I = \int_0^4 \int_0^{2\pi} (r \sin \theta)^2 r \, d\theta \, dr =$
$\int_0^4 \int_0^{2\pi} r^3 \sin^2\theta \, d\theta \, dr = \int_0^4 \int_0^{2\pi} [r^3(1 - \cos 2\theta)/2] \, d\theta \, dr =$
$\int_0^4 [r^3(\theta - \sin 2\theta/2)/2] \Big|_{\theta=0}^{2\pi} \, dr = \int_0^4 \pi r^3 \, dr = (\pi r^4/4) \Big|_0^4 = 64\pi$.

45. (a) By definition, the average value of a general function is the
volume divided by the area of the base. According to Exercise 26,
the volume of a linear function over a rectangle is equal to the
average of the heights of the four vertical edges times the area
of the base. Therefore, dividing this volume by the area of the
base implies that the average is just the average height of the
four vertices.

(b) A parallelopiped and a box with the same lengths, widths, and
heights have the same volumes. A parallelogram and a rectangle
with equal lengths and widths have equal areas. Therefore, in
analogy with the explanation in part (a), the average value of
a linear function on a parallelogram is equal to the average of
the heights of the four vertices.

49. From the theorem in Exercise 48 with $g(x,y) = f_y(x,y)$, we get
$(d/dx)\int_c^d f_y(x,y) \, dy = \int_c^d f_{xy}(x,y) \, dy$, i.e., $(d/dx)[f(x,d) - f(x,c)] =$
$\int_c^d f_{xy}(x,y) \, dy$. Thus, $[f(b,d) - f(b,c)] - [f(a,d) - f(a,c)] =$
$\int_a^b \int_c^d f_{xy}(x,y) \, dy \, dx$. The same argument with x and y interchanged
gives the same left-hand side and so $\int_a^b \int_c^d f_{xy}(x,y) \, dy \, dx =$
$\int_c^d \int_a^b f_{yx}(x,y) \, dx \, dy$, i.e., the double integrals of f_{xy} and f_{yx} are
the same. Since this can be done over any rectangle, f_{xy} and f_{yx}
must be the same. (One invokes here the fact that if the double integral
of a continuous function over arbitrarily small rectangles is 0 , then
the function is 0 .)

TEST FOR CHAPTER 17

1.   True or false.

   (a) $\int_0^\infty \exp(-x^2)dx = \sqrt{\pi}/2$.

   (b)   If $f(x,y) \geqslant 0$ for all x and y , and $b > a > 0$, $d > c > 0$,

         then $\int_a^b \int_c^d f(x,y)dy\ dx = \int_a^b \int_c^d \int_0^{f(x,y)} dz\ dy\ dx$.

   (c)   The mass of a solid W may be computed by using the triple

         integral $\iiint_W f(x,y,z)dx\ dy\ dz$, where $f(x,y,z)$ is the density

         of the solid at (x,y,z) .

   (d)   As long as f is integrable over W , it is always true that

         $\iiint_W f(x,y,z)dx\ dy\ dz = \iiint_W f(x,y,z)dz\ dx\ dy$.

   (e)   The geometric interpretation of the double integral $\iint_D dy\ dx$

         is either (i) the area of D or (ii) the volume of the region

         between the planes $z = 0$ and $z = 1$, and bounded by the boundary

         of D .

2.   Let A be the region bounded by the curve $y = x^4$ and the lines $y = |x^3|$

   (a)   Write the area of A as a double integral, integrating in y first.

   (b)   Rewrite the area of A by reversing the order of integration.

   (c)   Compute the area of A .

3.   (a)   Compute $\int_{-\infty}^\infty 2\pi \exp(-9x^2)dx$.

   (b)   Compute $\int_{-1}^1 \int_3^5 \int_0^{x^2} dy\ dx\ dz$ and interpret your answer geometrically.

4.   Find the average value of $\ln z$ over the region bounded by $z = 1$,

   $z = xy$, $x = 1$, $x = 3y$, $y = 2$, and $y = 3$.

5.   Find the volume between the cone $9 - x^2 - y^2$ and the xy-plane, where

   the domain is $x^2 + y^2 \leqslant 16$.

6.   Find the center of mass of the region described by $x \leqslant y \leqslant 3x$,

   $0 \leqslant x \leqslant 2$, and $-y \leqslant z \leqslant 0$.

7. Compute the volume which lies inside both the sphere $x^2 + y^2 + z^2 = 25$
    and the cylinder $y^2 + z^2 = 4$.

8. (a) Express the surface area of $f(x,y) = xy + 3$ over the rectangle
         $[1,3] \times [0,2]$ as a double integral. (Do not evaluate.)

    (b) Express the surface area of $g(x,y) = (x^2 + y^2)/2 + 8$ over the
         ellipse $x^2 + 4y^2 = 1$ as a double integral. (Do not evaluate.)

    (c) What is the relationship between the surface area of $f(x,y)$
         and $g(x,y)$ over the same domain?

9.

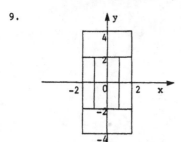

We wish to find the volume of the semi-ellipsoidal region described by $4x^2 + y^2 + 16z^2 = 16$ and $z \geqslant 0$. Use the grid shown at the left to make your estimates.

    (a) Use the maximum of z on each rectangle of the domain to compute
         an upper sum.

    (b) Use the minimum of z on each rectangle of the domain to compute
         a lower sum.

    (c) How is the actual volume V related to your answers in (a) and (b)?

    (d) Compute the actual volume to verify your answer in (c).

10. Recent research has shown that the number of additional strands of hair lost
    weekly is given by $\int_0^X \int_0^Y \sqrt{x}\, dy\, dx$, where X is the number of hours
    spent studying during the week and Y is the number of hours spent
    scratching parasites out of one's hair during the week. The theory is

10. (continued)

that studying causes the brain to bulge which causes hair roots to pop
out, and hair is pulled accidentally by scratching.

(a) Suppose a young college student spends 70 hours studying during
finals' week and 14 hours scratching bugs from unwashed hair.
How many additional strands of hair can this student expect to lose?

(b) Compare the answer in (a) to a normal week when studying occupies
12 hours and scratching occupies 3 hours.

ANSWERS TO CHAPTER TEST

1. (a) True

(b) True

(c) True

(d) True

(e) True

2. (a) $2\int_0^1 \int_{x^4}^{x^3} dy\, dx$

(b) $2\int_0^1 \int_{y^{1/3}}^{y^{1/4}} dx\, dy$

(c) 1/10

3. (a) $2\pi\sqrt{\pi}/3$

(b) 196/3 ; volume of region described by $0 \leqslant y \leqslant x^2$, $3 \leqslant x \leqslant 5$,
and $-1 \leqslant z \leqslant 1$.

4. (2007 ln 3 - 272 ln 2 - 1048)/523

5. 57π

6. $(3/8, 13/4, -13/8)$

7. $(500 - 84\sqrt{21})(\pi/3)$

8. (a) $\int_1^3 \int_0^2 \sqrt{1 + x^2 + y^2}\ dy\ dx$

    (b) $\displaystyle\int_{-1}^1 \int_{-\sqrt{1-4y^2}}^{\sqrt{1-4y^2}} \sqrt{1 + x^2 + y^2}\ dy\ dx$

    (c) Equal

9. (a) $12\sqrt{3} + 8$

    (b) $2\sqrt{2}$

    (c) $2\sqrt{2} \leqslant V \leqslant 12\sqrt{3} + 8$

    (d) 8π

10. (a) $1960\sqrt{70}/3$

    (b) $24\sqrt{12}$

CHAPTER 18

VECTOR ANALYSIS

18.1 Line Integrals

PREREQUISITES

1. Recall how to differentiate vector functions (Section 14.6).

2. Recall how to compute work for forces described in one variable (Section 9.5).

3. Recall how to describe a curve with parametric equations (Sections 2.4 and 10.4).

PREREQUISITE QUIZ

1. Compute $(d/dt)[t^3\underline{i} + (\cos t)\underline{j} - (e^t)\underline{k}]$.

2. Compute $(d/dx)[(\ln x)\underline{i} - (2x)\underline{j} + \underline{k}]$.

3. Calculate the work done by a force $F(x) = x^2$ if $0 \leqslant x \leqslant 2$.

4. Write a set of parametric equations to describe the unit circle.

5. Suppose that $y = 2x + 3$ and $y(t) = t - 3$, find x as a function of t .

GOALS

1. Be able to set up and compute a line integral and understand its relationship to work.

STUDY HINTS

1. __Line integrals and work__. Know that the line integral's physical inter-
   pretation is work. It is given by the formula $\int_{t_1}^{t_2} \Phi(\underline{\sigma}(t)) \cdot \underline{\sigma}'(t)dt$ or
   $\int_{t_1}^{t_2} \underline{F} \cdot \underline{v} \, dt$. Other notations that will be used include $\int_C \Phi(\underline{r}) \cdot d\underline{r}$,
   $\int_C \Phi$, and $\int_C (a \, dx + b \, dy + c \, dz)$.

2. __Computing work__. If you're given a vector field Φ and an oriented path
   $\underline{\sigma}$, computing the line integral (or work) is a simple matter of substitution. See Example 1. If a curve comes in pieces, one can integrate
   over each segment and add up the results from each segment. See Example
   8.

3. __Sign interpretation__. Positive work means that the force field did a net
   amount of work; i.e., the motion of the particle is in the direction of
   the force. If work is negative, then this amount of work is done by the
   particle on the force field.

4. __Parametrization of curves__. As long as a curve is parametrized with the
   correct direction and traces out the curve the correct number of times,
   the value of a line integral is independent of which parametrization is
   chosen. You can substitute for the endpoints to be sure the direction
   is correct. Example 5(a) shows that opposite directions yield opposite
   signs. Be sure the path is traversed the correct number of times.
   Example 5(b) shows how this can be a problem.

SOLUTIONS TO EVERY OTHER ODD EXERCISE

1. Use the formula $W = \int_{t_1}^{t_2} \Phi(\underline{\sigma}(t)) \cdot \underline{\sigma}'(t)dt$. We have $\underline{\sigma}(t) = 3t^2\underline{i} + t\underline{j} +$
   $\underline{k}$, $\underline{\sigma}'(t) = 6t\underline{i} + \underline{j}$, and $\Phi(\underline{\sigma}(t)) = 3t^2\underline{i} + t\underline{j}$. Thus, the work done
   is $\int_0^1 (3t^2\underline{i} + t\underline{j}) \cdot (6t\underline{i} + \underline{j})dt = \int_0^1 (18t^3 + t)dt = (9t^4/2 + t^2/2)\big|_0^1 = 5$.

5. At $(1,0,0)$, the kinetic energy, $K = 0$. At $(1,0,1)$, it is

    $(1/2)m\underline{v}\cdot\underline{v} = 5m/2 = 5/2$, since $m = 1$. Thus, the total work done is

    $5/2$. The work done by the force field is $\int_0^1 \underline{\Phi}(\underline{\sigma}(t))\cdot\underline{\sigma}'(t)dt$, where

    $\underline{\sigma}(t) = \underline{i} + t\underline{k}$, $\underline{\sigma}'(t) = \underline{k}$, and $\underline{\Phi}(\underline{\sigma}(t)) = \underline{k}$. Thus, $\int_0^1 (\underline{k}\cdot\underline{k})dt = 1$.

    Therefore, you did $3/2$ units of work.

9. By using the formula $W = \int_{t_1}^{t_2} \underline{\Phi}\ (\underline{\sigma}(t))\cdot\underline{\sigma}'(t)dt$, we have $\underline{\sigma}(t) =$

    $(\cos t, \sin t)$, $\underline{\sigma}'(t) = (-\sin t, \cos t)$, and $\underline{\Phi}(\underline{\sigma}(t)) = (-\sin t, \cos t)$.

    Thus, the work done is $\int_0^\pi (\sin^2 t + \cos^2 t)dt = \pi$.

13. The integral of $\underline{\Phi}$ along $\underline{\sigma}(t)$ for $t_1 \leqslant t \leqslant t_2$ is $\int_{t_1}^{t_2} \underline{\Phi}(\underline{\sigma}(t))\cdot\underline{\sigma}'(t)dt$.

    $\underline{\Phi}(\underline{\sigma}(t)) = \sin t\underline{i} + \cos t\underline{j} + t\underline{k}$; $\underline{\sigma}'(t) = \cos t\underline{i} - \sin t\underline{j} + \underline{k}$. Thus, the

    line integral is $\int_0^{2\pi} (\sin t\underline{i} + \cos t\underline{j} + t\underline{k})\cdot(\cos t\underline{i} - \sin t\underline{j} + \underline{k})dt =$

    $\int_0^{2\pi} t\ dt = (t^2/2)\big|_0^{2\pi} = 2\pi^2$.

17. The line integral of $\underline{\Phi}$ along $\underline{\sigma}(t)$ is $\int_{t_1}^{t_2} \underline{\Phi}(\underline{\sigma}(t))\cdot\underline{\sigma}'(t)dt$.

    (a) $\underline{\Phi}(\underline{\sigma}(t)) = \sin t\underline{i} + \cos t\underline{j} + \sin^3 t\underline{k}$; $\underline{\sigma}'(t) = \cos t\underline{i} + 2t\underline{j} + \underline{k}$,

    so the line integral is $\int_0^{2\pi}(\sin t \cos t + 2t \cos t + \sin^3 t)dt$.

    Let $u = \sin t$ to get $\int_0^{2\pi} \sin t \cos t\ dt = \int_0^0 u\ du = 0$. Let

    $u = \cos t$ to get $\int_0^{2\pi} \sin^3 t\ dt = \int_0^{2\pi} \sin t(1 - \cos^2 t)dt =$

    $-\int_1^1 (1 - u^2)du = 0$. Integrate by parts to get $\int_0^{2\pi} 2t \cos t\ dt =$

    $2t \sin t\big|_0^{2\pi} - \int_0^{2\pi} 2 \sin t\ dt = 0 + 2 \cos t\big|_0^{2\pi} = 0$. Therefore, the

    line integral is 0 .

21. The parametric form of the line is $\underline{\sigma}(t) = (t,t,t)$ for $0 \leqslant t \leqslant 1$;

    therefore, $\underline{\Phi}(\underline{\sigma}(t)) = t^2\underline{i} + t^2\underline{j} + \underline{k}$ and $\underline{\sigma}'(t) = \underline{i} + \underline{j} + \underline{k}$. Thus,

    the line integral is $\int_0^1 (t^2 - t^2 + 1)dt = t\big|_0^1 = 1$.

25. In this case, we have $dx = -3 \cos^2\theta \sin\theta$, $dy = 3 \sin^2\theta \cos\theta$; and

    $dz = 1$, and so, substituting $x = (\cos^3\theta)d\theta$, $y = (\sin^3\theta)d\theta$, and $z = \theta d\theta$,

    we get $\int_C [\sin z\ dx + \cos z\ dy - (xy)^{1/3}dz] = \int_0^{7\pi/2}(-3 \cos^2\theta \sin^2\theta +$

    $3 \sin^2\theta \cos^2\theta - \cos\theta \sin\theta)d\theta = -\int_0^{7\pi/2} \cos\theta \sin\theta\ d\theta =$

    $[(1/2)\sin^2\theta]\big|_0^{7\pi/2} = -1/2$.

29. $f(\underline{\sigma}(t)) = \sin t + \cos t + t \cos t$; $\underline{\sigma}'(t) = \cos t\underline{i} - \sin t\underline{j} + \underline{k}$, so

$\|\underline{\sigma}'(t)\| = \sqrt{\cos^2 t + (-\sin t)^2 + (1)^2} = \sqrt{2}$. Thus, the line integral is

$\sqrt{2}\int_0^{2\pi} (\sin t + \cos t + t \cos t)dt = \sqrt{2}[-\cos t\big|_0^{2\pi} + \sin t\big|_0^{2\pi} +$

$\int_0^{2\pi} t \cos t\, dt] = \sqrt{2}\int_0^{2\pi} t \cos t\, dt$. Integrate by parts with $u = t$

and $v = \sin t$: $\sqrt{2}[t \sin t\big|_0^{2\pi} - \int_0^{2\pi} \sin t\, dt] = \sqrt{2} \cos t\big|_0^{2\pi} = 0$.

33. $f(\underline{\sigma}(t)) = 6t^2$; $\underline{\sigma}'(t) = \underline{i} + 3\underline{j} + 2\underline{k}$, so $\|\underline{\sigma}'(t)\| =$

$\sqrt{(1)^2 + (3)^2 + (2)^2} = \sqrt{14}$. Thus, the line integral is $\sqrt{14}\int_1^3 6t^2 dt =$

$\sqrt{14}(2t^3)\big|_1^3 = 52\sqrt{14}$.

SECTION QUIZ

1.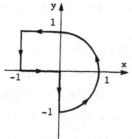

A particle traverses the "whistle" (composed of a semicircle and line segments) in a counterclockwise direction. The force acting on the particle is $\underline{F}(x,y) = 3\underline{i} + xy\underline{j}$.

(a) Compute the line integral of $\underline{F}$ along the "whistle" if it is traversed once starting at the origin.

(b) Interpret the sign and the physical meaning of your answer to part (a).

(c) Is the line integral different if the beginning and ending point is $(0,-1)$ rather than the origin? Explain.

(d) Is the line integral in part (a) different if the curve is traversed two times in a clockwise direction? Explain.

2. Compute $\int_C (x\, dx + y \sin z\, dy - e^{x/z}dz)$, where C is parametrized by $(x,y,z) = (t^2, \sin t, t)$, $0 \le t \le 2$.

3.   It was love at first sight for Handsome Harold and Lovely Lisa. On
     the evening of their first meeting, Cupid instantly shot his magic
     arrow through their hearts and united them. Witnesses say that the
     arrow travelled an unusual path: $\begin{cases} (t,0) & 0 \leq t \leq 2 \\ (1 - \cos(t-2+\pi), \sin(t-2+\pi)) & 2 \leq t \leq 2+\pi \end{cases}$.
     Suppose that a force field of $-10\underline{j}$ opposed the arrow. Was any work
     required to develop this romance, i.e., how much work was done by
     the magical love arrow?

ANSWERS TO PREREQUISITE QUIZ

1. $3t^2\underline{i} - (\sin t)\underline{j} - e^t\underline{k}$

2. $(1/x)\underline{i} - 2\underline{j}$

3. $8/3$

4. $x = \sin t$, $y = \cos t$, $0 \leq t \leq 2\pi$

5. $x = (t - 6)/2$

ANSWERS TO SECTION QUIZ

1.   (a) $1/2$

     (b) Work done by force field is $1/2$, work done by particle is $-1/2$.

     (c) No; it is just a reparametrization.

     (d) Yes; it is -1 .

2. $9 + (\sin^3 2)/3 - e^2$

3. 0

18.2 Path Independence

PREREQUISITES

1. Recall how to compute work by using line integrals (Section 18.1).

2. Recall how to compute a gradient vector (Section 16.1).

3. Recall how to compute partial derivatives (Section 15.1).

PREREQUISITE QUIZ

1. A force $xy\underline{i} + 3\underline{j}$ acts on a particle which moves from $(0,0)$ to

    $(1,1)$. How much work is done if the particle moves along the path:

    (a) $y = x$.

    (b) straight line segments from $(0,0)$ to $(1,0)$ and then to $(1,1)$.

2. If $f = 2 - 3xy$, what is the gradient of f ?

3. If $f(x,y,z) = 3x^2yz^{1/2}$, compute $\partial f/\partial x$, $\partial f/\partial y$, and $\partial f/\partial z$.

GOALS

1. Be able to determine if a vector field is conservative or not.

2. Be able to find the antiderivative of conservative vector fields.

STUDY HINTS

1. Conservative defined. A vector field is said to be conservative if its
    line integral doesn't depend on the path travelled, but only on the be-
    ginning and ending positions. Since the sign of the line integral
    changes with a change in the direction of movement, one conclusion
    that can be drawn is that the line integral of a conservative field
    around a closed curve is zero. The converse is not true; a single
    zero line integral does not imply that a vector field is conservative.
    (But if every line integral is zero, the field is conservative.)

2. <u>Gradients are conservative</u>. You must know this fact. If a vector field
    is a gradient, then the fact that its line integral depends only on the
    endpoints of the integration path is clear from the formula:
    $\int_C \nabla f(\underline{r}) \cdot d\underline{r} = f(B) - f(A)$, where A and B are the endpoints. The
    converse is also true. By definition, f is called the antiderivative.
    Only certain vector functions, namely gradients, have antiderivatives,
    since there may not be a single f that satisfies both $\partial f/\partial x = a$ and
    $\partial f/\partial y = b$ simultaneously.

3. <u>Cross-derivative test</u>. If a vector field $a(x,y)\underline{i} + b(x,y)\underline{j}$ is conser-
    vative, then $a = f_x$ and $b = f_y$ for some f . By the equality of
    mixed partials, we get $a_y = f_{xy} = f_{yx} = b_x$. Therefore, if a vector
    field is conservative, then $a_y = b_x$. The cross-derivative test on
    p. 898 asserts the converse. Knowing the little argument given for
    $a_y = b_x$ is useful to help you remember this test. Example 7 shows the
    importance of continuity. Exercise 30 extends this test to three vari-
    ables. Again, the equality of mixed partials is used.

4. <u>Finding antiderivatives</u>. You need to know how to find an antideriva-
    tive. Example 5 shows one method. Another method is to integrate each
    term and add on a constant which depends on the other variables. All
    of the antiderivatives should be equal. For example, suppose that the
    integrand is $4xyz\, dx + (2x^2 z + e^y)dy + (2x^2 y + 1)dz$. Integration in
    x yields $2x^2 yz + C(y,z)$. Integrating in y gives $2x^2 yz + e^y + C(x,z)$, and integrating in z gives $2x^2 yz + z + C(x,y)$. Comparing
    terms yields $f(x,y,z) = 2x^2 yz + e^y + z + C$. Note that $C(y,z) = e^y + z + C$; $C(x,z) = z + C$, and $C(x,y) = e^y + C$. Learn the method
    that is easiest for you.

SOLUTIONS TO EVERY OTHER ODD EXERCISE

1. The line integral of $\underline{\Phi}$ is path-independent because the vector field
    is conservative. The paths AOF and AODEF have the same endpoints,
    so the line integral is 3 .

5. A nonzero line integral around a closed curve shows that the vector field
    is not conservative. Let C consist of three paths: C_1 is $(0,t,0)$,
    $0 \leqslant t \leqslant 1$; C_2 is $(t - 1,1,0)$, $1 \leqslant t \leqslant 2$; C_3 is $(3 - t,3 - t,0)$,
    $2 \leqslant t \leqslant 3$. On C_1 , C_2 , and C_3 , $\underline{\sigma}'(t)$ is $(0,1,0)$, $(1,0,0)$,
    and $(-1,-1,0)$, respectively. Thus, $\int_C \underline{\Phi}(\underline{\sigma}(t)) \cdot \underline{\sigma}'(t)dt =$
    $\int_0^1 (t,t,1) \cdot (0,1,0)dt + \int_1^2 (1,1,1) \cdot (1,0,0)dt + \int_2^3 (3 - t,3 - t,1) \cdot$
    $(-1,-1,0)dt = \int_0^1 t\,dt + \int_1^2 dt + \int_2^3 (2t - 6)dt = (t^2/2)\big|_0^1 + t\big|_1^2 +$
    $(t^2 - 6t)\big|_2^3 = 1/2$. Thus, $\underline{\Phi}$ is not conservative.

9. We use the fact that if $\underline{\Phi}$ is a gradient of $f(x,y)$, then $\int_C \underline{\Phi} =$
    $f(B) - f(A)$, where A and B are the endpoints of C . We recognize
    $2xy\,dx + x^2 dy$ as the gradient of $x^2 y$. $A = (1,0)$ and $B = (0,5\sqrt{2}/2)$,
    so the line integral is $f(B) - f(A) = 0$.

13. $\underline{F} = -[JMm/(x^2 + y^2 + z^2)^{5/2}](x\underline{i} + y\underline{j} + z\underline{k})$, so we guess this to be the
    gradient of $JMm(x^2 + y^2 + z^2)^{-3/2}$. Differentiation shows that, in
    fact, it is the gradient of $JMm(x^2 + y^2 + z^2)^{-3/2}/3 = JMmr^{-3}/3$. Since
    the force field is a gradient, the work or line integral, depends only
    on the endpoints; it is $f(B) - f(A) = JMm(r_2^{-3} - r_1^{-3})/3$.

17. We use the cross-derivative test to determine if $a(x,y)\underline{i} + b(x,y)\underline{j}$ is
    conservative; it is if $a_y = b_x$. If the vector is conservative, use
    the method of Example 5 to find an antiderivative. $a_y = 2x$ and $b_x =$
    $2x$. Since $a_y = b_x$, the vector is conservative. We integrate
    $a(x,y) = 2xy$ with respect to x to get $x^2 y + g(y)$. Differentiation
    with respect to y gives $x^2 + g'(y)$, and comparison with $b(x,y)$

17. (continued)

gives $g'(y) = \cos y$. Therefore, $g(y) = \sin y$, so the antiderivative is $f(x,y) = x^2 y + \sin y + C$.

21. If $f(x,y) = \tan^{-1}(y/x)$, then $\partial f/\partial x = (-y/x^2)[1/(1 + (y/x)^2] = (-y/x^2)x^2/(x^2 + y^2) = -y/(x^2 + y^2)$. Also, $\partial f/\partial y = (1/x)[1/(1 + (y/x)^2)] = (1/x)x^2/(x^2 + y^2) = x/(x^2 + y^2)$. Thus, $\underline{\nabla}f$ is the vector field of Example 7.

25. We want a function f such that $\partial f/\partial x = x$, $\partial f/\partial y = y$, and $\partial f/\partial z = z$. Such a function is $(x^2 + y^2 + z^2)/2$. The work done is $f(\underline{\sigma}(b)) - f(\underline{\sigma}(a))$. $\underline{\sigma}(0) = (1,0,0)$ and $\underline{\sigma}(\pi) = (0,1,0)$; therefore, the work done is $(1/2) - (1/2) = 0$.

29. We are integrating over a gradient, so $\int_C \underline{\nabla}f(\underline{r})\cdot d\underline{r} = f(\underline{\sigma}(1)) - f(\underline{\sigma}(0)) = f(1/2,\sqrt{2}/2,3) - f(0,0,2) = (1/2)^3 - (\sqrt{2}/2)^3 + \sin(3\sqrt{2}\pi/4)$.

33. $C_y = x \exp(yz) + xyz \exp(yz) = b_z$; $c_x = y \exp(yz) = a_z$; $a_y = z \exp(yz) = b_x$. Since the three equalities are satisfied, the field must be conservative. Integrating a , b , c with respect to x , y , and z , respectively gives $x \exp(yz) + C(y,z)$, $x \exp(yz) + C(x,z)$, and $x \exp(yz) + C(x,y)$, where $C(y,z)$ is a function depending only on y and z , etc. Comparison shows that the antiderivative is $f(x,y,z) = x \exp(yz) + C$.

37. A closed curve crosses the union of circles an even number of times because, if a curve "enters" a circle, it must also "leave" the circle; therefore, the number of intersections is a multiple of two.

Pick an arbitrary fixed point P . If an arc from P to a certain region crosses the circles an even number of times, color the region red.

If the number is odd, color the region blue. Notice that a single

37. (continued)

crossing takes one from a region to an adjacent region, so adjacent regions have different colors. To see that this coloring scheme is consistent, we must verify that, if $\underline{\sigma}_1$ and $\underline{\sigma}_2$ are arcs from P to a given region, then the numbers n_1 and n_2 of crossings are both even or both odd. We must show that $n_1 - n_2$ is even, or, equivalently, that $n_1 + n_2 = n_1 - n_2 + 2n_2$ is even. But $n_1 + n_2$ is the number of crossings for a closed curve that goes along $\underline{\sigma}_1$ from P to the region and then back along $\underline{\sigma}_2$ to P. By our first observation, this number of crossings is even.

SECTION QUIZ

1. Let $f(x,y) = x^3 + \cos y$. What is $\int_C \underline{\nabla} f(\underline{r}) \cdot d\underline{r}$, where C is the path along the unit circle, which is traversed $2\frac{1}{2}$ times from $(0,-1)$ to $(0,1)$? If this can't be done, explain what is missing.

2. Are the following vector fields conservative? If yes, find an antiderivative:

   (a) $(x^2 y + \cos y)\underline{i} + (x^3/3 - x \sin y)\underline{j}$

   (b) $(3 + e^x)\underline{i} - y\underline{j}$

   (c) $[\exp(xy) + y \cos x]\underline{i} + [\exp(xy) + x \cos y]\underline{j}$

   (d) $(1/y)\underline{i} - (x/y^2)\underline{j}$

3. True or false: If $\int_C \underline{\Phi} \cdot d\underline{r} = 0$, then $\underline{\Phi}$ is a conservative vector field.

4. Once again, Tardy Terry arrived late for work. She claims that she oversleeps a lot because she gets tired out from sleepwalking. Last night, a howling wind came through an open window with force $\underline{F} =$

4.   (continued)

$(2xye^x + x^2ye^x)\underline{i} + x^2e^x\underline{j}$ and swept her through the house while she sleepwalked.

(a)   Is the work done by the wind independent of path? Explain.

(b)   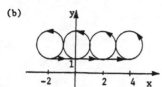 At one time last night, Tardy Terry walked the path shown at the left. Each circle of radius one was traversed once in the specified direction. Compute the line integral along the path from (-2,1) to (4,1) .

ANSWERS TO PREREQUISITE QUIZ

1.   (a) $10/3$

(b) 3

2. $-3y\underline{i} - 3x\underline{j}$

3. $\partial f/\partial x = 6xyz^{1/2}$; $\partial f/\partial y = 3x^2z^{1/2}$; $\partial f/\partial z = 3x^2y/2z^{1/2}$

ANSWERS TO SECTION QUIZ

1. $2\cos(1)$; orientation isn't necessary here since the line integral is independent of path.

2.   (a) $x^3y/3 + x\cos y + C$

(b) $3x + e^x - y^2/2 + C$

(c)   Not conservative

(d)   Not conservative; vector field is undefined if $y = 0$.

3.   False; let $\underline{\Phi} = y\underline{i}$ and let C be the line segment from (0,0) to (0,1) to (0,0)

4.   (a)   Yes; the cross-derivative test is satisfied.

(b) $16e^4 - 4e^{-2}$

18.3 Exact Differentials

PREREQUISITES

1.   Recall the cross-derivative test (Section 18.2).

2.   Recall how to solve separable differential equations in one variable
     (Section 8.5).

PREREQUISITE QUIZ

1.   State the cross-derivative test.

2.   Is $\underline{\phi} = xy\underline{i} + xy\underline{j}$ conservative? Explain.

3.   Solve the differential equation $dy/dx = (x^2 + x)/y$.

GOALS

1.   Be able to solve exact differential equations.

2.   Be able to use integrating factors to transform differential equations
     into exact ones.

STUDY HINTS

1.   Cross-derivative test. Only the notation is different from the test
     presented in Section 18.2. P replaces $a(x,y)$ and Q replaces $b(x,y)$.
     Again, if the mixed partial derivatives are equal, the vector field is a
     gradient. If they are not equal, no antiderivative exists.

2.   Finding an antiderivative. Yet another method for finding antiderivatives
     is discussed in method 2 of Example 1. Another method was given in Example
     5 of Section 18.2, and another one was discussed in the study hints of the
     last section. All of these methods yield the same results. Learn the one
     that is easiest for you.

3. **Exact differentials.** This is just another term to describe an integrand which has previously been described by the words "conservative" and "gradient".

4. **Exact differential equations.** $P + Q(dy/dx) = 0$ can be multiplied by dx to get $Pdx + Qdy = 0$. If $Pdx + Qdy$ is exact, and $P = f_x$ and $Q = f_y$, then $f(x,y) = C$ is the solution of $P + Q(dy/dx) = 0$. This is because $df = (\partial f/\partial x)dx + (\partial f/\partial y)dy$ and $df = 0$ means that f is constant.

5. **Integrating factors.** This is a device used to transform equations into exact ones. If we multiply $M + N(dy/dx) = 0$ by an integrating factor, μ, we get $\mu M + \mu N(dy/dx) = 0$. Taking partials, we want $(\mu M)_y = (\mu N)_x$. According to the text, this is possible if $(M_y - N_x)/N$ is a factor of x alone in which case, $\mu = \exp[\int((M_y - N_x)/N)dx]$ _if_ μ is a function of x alone. It is unpleasant and impractical to memorize this. Instead, try multiplying by $\mu(x)$ and apply the cross-derivative test to find μ. If this fails, try multiplying by $\mu(y)$.

SOLUTIONS TO EVERY OTHER ODD EXERCISE

1. Let $P = 2x + (xy + 1)\exp(xy)$ and $Q = x^2\exp(xy)$. Then, $\partial P/\partial y = x \exp(xy) + (xy + 1)y \exp(xy)$ and $\partial Q/\partial x = 2x \exp(xy) + x^2(x) \exp(xy)$. Since $\partial P/\partial y \neq \partial Q/\partial x$, the function is not an exact differential.

5. Use the cross-derivative test. Let $P = x^2y$ and $Q = x^3y/3$. Then $\partial P/\partial y = x^2$ and $\partial Q/\partial x = 3x^2y/3 = x^2y$. Since $\partial P/\partial y \neq \partial Q/\partial x$, the expression is not exact.

9. The function f is the integral of $Pdx + Qdy$ along any path from $(0,0)$ to (x,y). $\hat{f}(x,y)$ is the integral along the path from $(0,0)$ to $(0,y)$ and then from $(0,y)$ to (x,y).

13. If $\partial P/\partial y = \partial Q/\partial x$ for the equation $P + Q(dy/dx) = 0$, then the dif-
    ferential equation is exact and can be solved by the method of Example
    3. $P = ye^x + e^y$, so $\partial P/\partial y = e^x + e^y$; $Q = xe^y + e^x$, so $\partial Q/\partial x = e^y +$
    e^x . It is exact, so $f(x,y) = \int_0^x dt + \int_0^y (xe^t + e^x)dt = t\big|_0^x + (xe^t +$
    $te^x)\big|_0^y = x + xe^y + ye^x - x = xe^y + ye^x = C$. $y(0) = 2$ implies $C = 2$;
    therefore, the solution is $xe^y + ye^x = 2$.

17. $P + Q(dy/dx) = 0$ is exact if $\partial P/\partial y = \partial Q/\partial x$. Then, the equation is
    solved by the method of Worked Example 3. $P = y + x^2 y^2 - 1$, so
    $\partial P/\partial y = 1 + x^2$; $Q = x$, so $\partial Q/\partial x = 1$. The differential equation
    is not exact.

21. Use the method of Example 4.

    (a) $\partial P/\partial y = -2x(2y)/(y^2 + 1)^2 = -4xy/(y^2 + 1)^2$ and $\partial Q/\partial x = -4xy/$
        $(y^2 + 1)^2$. The integrand is exact, so we may reparametrize the
        curve from $(-1,0)$ to $(0,0)$. Let it be $(t - 1,0)$, $0 \leqslant t \leqslant 1$.
        Then the line integral is $\int_0^1 2(t - 1)dt = 2(t^2/2 - t)\big|_0^1 = -1$.

    (b) Rearrangement of the integrand in (a) yields $x/y = [(x^2 + 1)/$
        $(y^2 + 1)]\,dy/dx$. The antiderivative of the integrand is
        $(x^2 + 1)/(y^2 + 1)$, so the solution of the differential equation
        is $(x^2 + 1)/(y^2 + 1) = C$. $y(1) = 1$ implies $C = 1$, so the
        solution is $(x^2 + 1)/(y^2 + 1) = 1$, i.e., $x^2 + 1 = y^2 + 1$,
        i.e., $x = y$. Since $y(1) = 1$, $y = -x$ is not a solution.

25. $\mu = \exp\left[\int [(M_y - N_x)/N]\,dx\right]$. $M = 2y \cos y + x$, so $M_y = 2 \cos y -$
    $2y \sin y$; $N = x \cos y - xy \sin y$, so $N_x = \cos y - y \sin y$. Thus,
    $(M_y - N_x)/N = (\cos y - y \sin y)/x(\cos y - y \sin y) = 1/x$. Therefore,
    $\mu = \exp(\int (1/x)dx) = \exp(\ln(x)) = x$.

29. If μ is a function of y alone, then $\mu_y = \mu'(y)$ and $\mu_x = 0$.
Then, $\mu_y M + \mu M_y = \mu_x N + \mu N_x$ becomes $\mu_y M = \mu(N_x - M_y)$ or $\mu'/\mu = (N_x - M_y)/M$, i.e., $\ln \mu = \int [(N_x - M_y)/M] \, dy$.

SECTION QUIZ

1. Determine which of the following differential equations is exact, and solve the ones which are exact.

    (a) $(5xy + 3 + \sin y) + (5x^2/2 + x \cos y) dy/dx = 0$; $y(0) = 2$.

    (b) $(\sin xy + 2x + 5y) + (\sin xy + 2x + 5y) dy/dx = 0$; $y(0) = 1$.

    (c) $(ye^x - \sinh y) + (e^x - x \cosh y)y' = 0$; $y(0) = -1$.

2. Use the method of integrating factors to solve the differential equation $x^2 y + x^3 (dy/dx) = 0$.

3. At Andy the Anteater's favorite anthole, only the fastest ants dare to venture away from the hole. Andy's tongue, known as the fastest in the West, flicks out at a speed v , which is dependent upon his position r from the anthole. It is known that v and r are related by $v^2 + ve^{vr} + (2rv + re^{vr})(dv/dr) = 0$. Knowing that $v(0) = 4$, solve the differential equation.

ANSWERS TO PREREQUISITE QUIZ

1. A vector field $\underline{\Phi}(x,y) = a(x,y)\underline{i} + b(x,y)\underline{j}$ is conservative if and only if $\partial a/\partial y = \partial b/\partial x$.

2. No; $a_y = x \neq y = b_x$.

3. $y^2/2 = x^3/3 + x^2/2 + C$

ANSWERS TO SECTION QUIZ

1. (a) $5x^2y/2 + x \sin y + 3x + 2 = 0$

    (b) Not exact

    (c) $ye^x - x \sinh y + 1 = 0$

2. $xy = C$

3. $rv^2 + e^{vr} = 1$

18.4 Green's Theorem

PREREQUISITES

1. Recall how to compute line integrals (Section 18.1).

2. Recall how to perform double integration (Sections 17.1, 17.2, and 17.3).

3. Recall how to parametrize a curve (Sections 2.4 and 10.4).

PREREQUISITE QUIZ

1. If $\underline{\phi} = z\underline{i} - 2yx\underline{j}$ and C is the following curve: $(x,y,z) = (1 - t, t, t^2)$, $0 \leqslant t \leqslant 2$, what is $\int_C \underline{\phi} \, dt$?

2. Compute $\iint_D \cos y \, dx \, dy$ where D is the square $[0,\pi/2] \times [\pi,3\pi/2]$.

3. 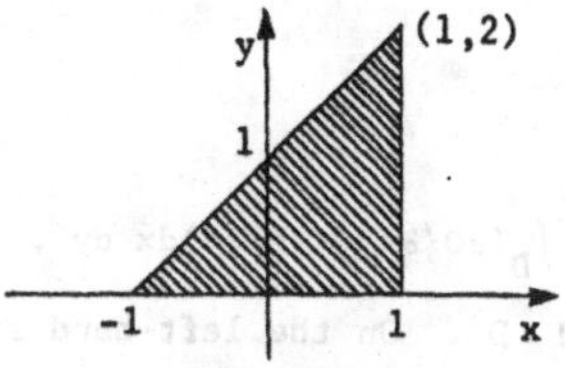 Write the area of the triangle as a double integral.

4. Given that $x = a \sin bt$, $0 \leqslant t \leqslant 2\pi$; find the values for a and b and find an expression for y such $x^2 + y^2 = 4$.

GOALS

1. Be able to convert certain line integrals to double integrals and vice versa by using Green's theorem.

2. Be able to calculate areas by using Green's theorem.

STUDY HINTS

1. Green's theorem. This theorem allows you to convert line integrals to double integrals, which may be easier to compute. You should memorize the result $\int_C (Pdx + Qdy) = \iint_D (\partial Q/\partial x - \partial P/\partial y)dx \, dy$, where C is the

1.   (continued)

   boundary of D . Example 4 shows one of the advantages of Green's theorem.

2.   **Key points.** Notice that C is closed and must be traversed in a counter-

   clockwise direction. Also continuous partial derivatives are required.

   Example 1, Section 18.5 shows what happens if the conditions are not met.

   If Green's theorem does not apply directly to a region, the region may be

   subdivided so that the theorem can be applied. See Example 2.

3.   **Area.** From Green's theorem, we get $A = (1/2)\int_C (xdy - ydx)$. In most

   cases, you will compute areas by using normal methods of integration.

   However, if the curve bounding the region is given parametrically, Green's

   theorem can be <u>much</u> easier; a case is Exercise 24, p. 913.

SOLUTIONS TO EVERY OTHER ODD EXERCISE

1.   We want to show that $\int_C (Pdx + Qdy) = \iint_D (\partial Q/\partial x - \partial P/\partial y)dx\,dy$, where

   C is traversed counterclockwise around D . On the left-hand side, we

   have $\int_C (Pdx + Qdy) = \int_C (xy\,dx + xdy)$. x and x^2 intersect at x = 0

   and x = 1 , so let C_1 be (t, t^2) for $0 \leqslant t \leqslant 1$ and let C_2 be

   (1 - t, 1 - t) for $0 \leqslant t \leqslant 1$. Thus, the line integral is

   $\int_0^1 [(t)(t^2)(dt) + t(2t\,dt)] + \int_0^1 [(1 - t)(1 - t)(-dt) + (1 - t)(dt)] =$

   $\int_0^1 (t^3 + 2t^2 - (1 - 2t + t^2) - (1 - t))dt = \int_0^1 (t^3 + t^2 + 3t - 2)dt =$

   $(t^4/4 + t^3/3 + 3t^2/2 - 2t)\big|_0^1 = 1/12$.

     The right-hand side is $\iint_D (\partial Q/\partial x - \partial P/\partial y)dx\,dy = \int_0^1 \int_{x^2}^x (1 - x)dy\,dx =$

   $\int_0^1 [(y - xy)\big|_{y=x^2}^x]dx = \int_0^1 (x^3 - 2x^2 + x)dx = (x^4/4 - 2x^3/3 + x^2/2)\big|_0^1 = 1/12$

5.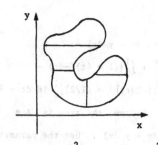

The sketch at the left shows only one way of dividing the region into type 1 and type 2 subregions.

9. Let $P = xy^2$ and $Q = -yx^2$; thus, $\partial Q/\partial x = -2xy$ and $\partial P/\partial y = 2xy$.

By Green's theorem, $\int_C \underline{\phi}(\underline{r}) \cdot d\underline{r} = \iint_D - 4xy \, dx \, dy =$

$-4\int_{-b}^{b} \int_{-(a/b)\sqrt{b^2-y^2}}^{(a/b)\sqrt{b^2-y^2}} xy \, dx \, dy = -4\int_{-b}^{b} [(x^2 y/2)\Big|_{x=-(a/b)\sqrt{b^2-y^2}}^{(a/b)\sqrt{b^2-y^2}}] dy =$

$-4\int_{-b}^{b} 0 \, dy = 0$.

13. Using Green's theorem, we have $\int_C (P \, dx + Q \, dy) = \iint_D (\partial Q/\partial x - \partial P/\partial y) dx \, dy =$
$\int_1^3 \int_2^3 (\partial Q/\partial x - \partial P/\partial y) dy \, dx$. Here, $\partial Q/\partial x = 0$ and $\partial P/\partial y = 4y$, so the integral is $\int_1^3 \int_2^3 (-4y) dy \, dx = \int_1^3 (-2y^2)\Big|_{y=2}^{3} dx = -10\int_1^3 dx = -20$.

17. (a) Recall that a zero dot product implies that the vectors are orthogonal. $(P\underline{i} + Q\underline{j}) \cdot (Q\underline{i} - P\underline{j}) = PQ - QP = 0$.

(b) By Green's theorem, we have $\iint_D (\partial P/\partial x - \partial(-Q)/\partial y) dx \, dy =$ $\int_C [(-Q)dx + P \, dy]$. Given some parametrization of C : $x = x(t)$, $y = y(t)$, we know that $dx = kP$ and $dy = kQ$, where k is a constant, because $P\underline{i} + Q\underline{j}$ is parallel to the tangent vector. Therefore, the line integral is $\int_C [(-Q)(kP) + P(kQ)] dt = \int_C 0 \, dt = 0$.

21.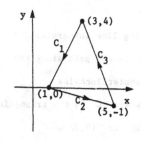

Let C be composed of C_1 , C_2 , and C_3 as shown at the left. For $0 \leqslant t \leqslant 1$, C_1 is $(1 - t)(3,4) + t(1,0) = (3 - 2t, 4 - 4t)$, C_2 is $(1 - t)(1,0) + t(5,-1) = (1 + 4t,-t)$, and C_3 is $(1 - t)(5,-1) + t(3,4) = (5 - 2t,-1 + 5t)$. By the corollary,

21. (continued)

the area of the triangle is $A = (1/2)\int_C (x \, dy - y \, dx) =$

$(1/2)\{\int_0^1 [(3 - 2t)(-4dt) - (4 - 4t)(-2dt)] + \int_0^1 [(1 + 4t)(-dt) -$

$(-t)(4dt)] + \int_0^1 [(5 - 2t)(5dt) - (-1 + 5t)(-2dt)]\} = (1/2)\int_0^1 18 \, dt = 9$.

25.

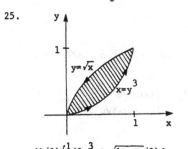

By Green's theorem, the area is $A = (1/2)\int_C (x \, dy - y \, dx)$. Use the parametrization (t, t^3) , $0 \leqslant t \leqslant 1$ and $(1 - t, \sqrt{1 - t})$, $0 \leqslant t \leqslant 1$. Thus, $A = (1/2)\{\int_0^1 [(t)(3t^2 dt) - (t^3)(dt)] +$

$\int_0^1 [(1 - t)(-dt/2\sqrt{1 - t}) - (\sqrt{1 - t})(-dt)]\} =$

$(1/2)\int_0^1 (2t^3 + \sqrt{1 - t}/2)dt = (1/2)[t^4/2 - (1 - t)^{3/2}/3]\big|_0^1 = (1/2)(1/2 +$

$(1/3) = 5/12$.

29. $\partial P/\partial y = 1 = \partial Q/\partial x$,   so $P \, dx + Q \, dy$ is an exact differential; therefore, it is conservative and the line integral is independent of path. A closed curve begins and ends at the same point, so $\int_C (P \, dx + Q \, dy) = 0$.

33. As is described in most encyclopedias, a planimeter is a mechanical device for measuring the area of a region. The device is run around the boundary of the region while a simple, but a clever mechanism actually performs the integration $(1/2)\int_{\partial D} (x \, dy - y \, dx)$ mechanically, during the circuit. This integral is the area of the region, by Green's theorem.

SECTION QUIZ

1. Use Green's theorem to compute the following line integrals:

(a) $\int_C (5xy^2 \, dx + y \sin x \, dy)$, where   C   is the path along the square $[0,\pi] \times [0,\pi]$, traversed counterclockwise.

(b) $\int_C [(5x^2 + 6y^3)dx + (e^x + 4xy)dy]$, where   C   is the triangular path from   (0,0)   to   (0,1)   to   (1,0)   to   (0,0) .

2. Use Green's theorem to find the area of the region bounded by the

    parametrized curve (t, t^2), $-1 \leqslant t \leqslant 2$ and the line segment joining

    the endpoints.

3. A bearded old man, dressed in rags, wishes to know about his future.

    A fortune teller informs him that the crystal ball shows a harem of

    dancing girls inside an odd shaped room bounded by the ellipse

    $x^2 + y^2/4 = 1$.

    (a) Give a parametrization for the boundary of the room.

    (b) Use the corollary to Green's theorem to calculate the area of

         the room.

    (c) If the average harem girl occupies $(1/8)$ square units, how

         many girls can the dirty (according to his clothing) old man keep?

ANSWERS TO PREREQUISITE QUIZ

1. $-4/3$

2. -2π

3. $\int_{-1}^{1} \int_{0}^{x+1} dy \, dx$

4. $a = 2$; $b = 1$; $y = 2 \cos t$

ANSWERS TO SECTION QUIZ

1. (a) $-5\pi^4/2$

    (b) $17/6 - e$ (Note the orientation of C .)

2. $9/2$

3. (a) $(\cos t, 2 \sin t)$, $0 \leqslant t \leqslant 2\pi$

    (b) 2π

    (c) 16π or about 49 girls

18.5 Circulation and Stokes' Theorem

PREREQUISITES

1.   Recall how to use Green's theorem (Section 18.4).

2.   Recall how to compute a cross product (Section 13.5).

PREREQUISITE QUIZ

1.   When using Green's theorem, in what direction must the boundary be
     traversed?

2.   Fill in the blanks: Green's theorem states that $\int_C$ _____ =
     $\iint_D$ _____ .

3.   Use Green's theorem to evaluate $\int_C [(x^2 y + y)dx + (3 \cos x)dy]$,
     where C is the rectangle $[0,1] \times [2,3]$ traversed clockwise.

4.   Evaluate $(3\underline{i} + 2\underline{j} + \underline{k}) \times (-\underline{i} - 2\underline{j} + 2\underline{k})$.

GOALS

1.   Be able to compute the curl of a vector field and explain its physical
     interpretation.

2.   Be able to convert a line integral into a surface integral by using
     Stokes' theorem.

STUDY HINTS

1.   Scalar curl.   For $\underline{\phi} = P\underline{i} + Q\underline{j}$, the scalar curl is defined to be
     $\partial Q/\partial x - \partial P/\partial y$. This is the integrand of the double integral in
     Green's theorem. Physically, the scalar curl tells you about the
     circulation per unit area. Circulation is a term which describes
     the rotational motion of fluids, such as tea circulating in a cup or air
     circulating in a room.

2. **Surface integral.** This is defined as $\iint_S \phi \cdot \underline{n} dA = \iint_S \underline{\phi} \cdot d\underline{A} =$
    $\iint_D (P\underline{i} + Q\underline{j} + R\underline{k}) \cdot (-f_x \underline{i} - f_y \underline{j} + \underline{k}) dx\, dy$, if S is the graph of f .

3. **Curl.** It is defined by curl $\underline{\phi} = (R_y - Q_z)\underline{i} + (P_z - R_x)\underline{j} + (Q_x - P_y)\underline{k}$.
    You should remember this by curl $\underline{\phi} = \underline{\nabla} \times \underline{\phi}$, where $\underline{\nabla} = (\partial/\partial x)\underline{i} +$
    $(\partial/\partial y)\underline{j} + (\partial/\partial z)\underline{k}$.

4. **Stokes' theorem.** Green's theorem was restricted to the xy-plane.
    Stokes' theorem, like Green's, allows you to convert a line integral
    to a double integral. It is more general in that the boundary, ∂S ,
    and the region of integration, S , are no longer restricted to a
    plane. Again, the boundary must be traversed counterclockwise and
    continuous differentiability is required. By "counterclockwise," we
    mean that if you walk around the boundary, the region is on your left
    (See Fig. 18.5.4). The formula is $\int_{\partial S} \underline{\phi}(\underline{r}) \cdot d\underline{r} = \iint_S (\underline{\nabla} \times \underline{\phi}) \cdot \underline{n} dA$. Note
    that if $\underline{\phi} = P(x,y)\underline{i} + Q(x,y)\underline{j}$ and S is a plane region, you get
    Green's theorem. (Some students like to use this method to remember
    Green's theorem.)

5. **Applications.** Method 2 of Example 7 shows one of the uses of Stokes'
    theorem. According to the theorem, we can change a surface to any other
    surface with the same boundary. In most cases, we will change to a
    planar surface.

6. **Example 7 clarified.** Some people have said that $\underline{m}$ stands for
    "mysterious" in Method 1 of Example 7.
    Looking at the plane x + y + z = 1 edge
    on, we see that it is a circle that is cut
    out and that the vector required is ortho-
    gonal to the plane with length equal to the distance of the plane from
    the origin. Thus, $\underline{m} = a(\underline{i} + \underline{j} + \underline{k})$. Using the distance from a point

7.   (continued)

to a plane equation (formula (7), p. 674), we get $\|\underline{m}\| = 1/\sqrt{3}$, so

$a = 1/3$. Therefore, $\underline{m} = (1/3)(\underline{i} + \underline{j} + \underline{k})$.

SOLUTIONS TO EVERY OTHER ODD EXERCISE

1. $\partial Q/\partial x - \partial P/\partial y$ is the scalar curl of the vector field $P\underline{i} + Q\underline{j}$.

$\partial Q/\partial x = -1$ and $\partial P/\partial y = 1$, so the scalar curl is -2 .

5.   The surface integral of $\underline{\Phi}$ over S , where $\underline{\Phi} = P\underline{i} + Q\underline{j} + R\underline{k}$ and S

is the surface $z = f(x,y)$, is $\iint_D(-Pf_x - Qf_y + R)dx\,dy$. $z = 2x -$

y , so $f_x = 2$ and $f_y = 1$; $P = 3x^2$, $Q = -2yx$, and $R = 8$.

Thus, the surface integral is $\int_0^2\int_0^2(-6x^2 + 2xy + 8)dx\,dy = \int_0^2(-2x^3 +$

$x^2 y + 8x)\big|_{x=0}^2 dy = \int_0^2 4y\,dy = 2y^2\big|_0^2 = 8$.

9.   The curl of $\underline{\Phi} = P\underline{i} + Q\underline{j} + R\underline{k}$ is $(R_y - Q_z)\underline{i} + (P_z - R_x)\underline{j} + (Q_x - P_y)\underline{k} =$

$\underline{\nabla} \times \underline{\Phi}$. In this case, $\underline{\nabla} \times \underline{F} = \begin{vmatrix} \underline{i} & \underline{j} & \underline{k} \\ \partial/\partial x & \partial/\partial y & \partial/\partial z \\ e^z & -\cos(xy) & z^3 y \end{vmatrix} = z^3\underline{i} + e^z\underline{j} +$

$(y \sin xy)\underline{k}$.

13.   Curl $(f\underline{\Phi}) = \begin{vmatrix} \underline{i} & \underline{j} & \underline{k} \\ \dfrac{\partial}{\partial x} & \dfrac{\partial}{\partial y} & \dfrac{\partial}{\partial z} \\ fP & fQ & fR \end{vmatrix}$, where $\underline{\Phi} = P\underline{i} + Q\underline{j} + R\underline{k}$. This is

$(\partial fR/\partial y - \partial fQ/\partial z)\underline{i} + (\partial fP/\partial z - \partial fR/\partial x)\underline{j} + (\partial fQ/\partial x - \partial fP/\partial y)\underline{k}$. By

the product rule for differentiation, we get $[(\partial f/\partial y)R + f(\partial R/\partial y) -$

$(\partial f/\partial z)Q - f(\partial Q/\partial z)]\underline{i} + [(\partial f/\partial z)P + f(\partial P/\partial z) - (\partial f/\partial x)R - f(\partial R/\partial x)]\underline{j} +$

$[(\partial f/\partial x)Q + f(\partial Q/\partial x) - (\partial f/\partial y)P - f(\partial P/\partial y)]\underline{k} = f[(R_y - Q_z)\underline{i} +$

$(P_z - R_x)\underline{j} + (Q_x - P_y)\underline{k}] + [(\partial f/\partial y)R - (\partial f/\partial z)Q]\underline{i} + [(\partial f/\partial z)P -$

$(\partial f/\partial x)R]\underline{j} + [(\partial f/\partial x)Q - (\partial f/\partial y)P]\underline{k}$. We recognize the first half as

f curl $\underline{\Phi}$ and the second half is $\underline{\nabla}f \times \underline{\Phi}$.

17. $\underline{\nabla} \times \underline{\Phi} = \begin{vmatrix} \underline{i} & \underline{j} & \underline{k} \\ \partial/\partial x & \partial/\partial y & \partial/\partial z \\ P & Q & Q \end{vmatrix}$, where $P = 1/(y + z)$ and $Q =$

$-x/(y + z)^2$. By symmetry, note that $\partial Q/\partial y = \partial Q/\partial z$ and $\partial P/\partial y = \partial P/\partial z$. Thus, $\underline{\nabla} \times \underline{\Phi}$ reduces to $(\partial Q/\partial x - \partial P/\partial y)\underline{k}$. However, $\partial Q/\partial x = -1/(y + z)^2$ and $\partial P/\partial y = -1/(y + z)^2$; therefore, $\underline{\nabla} \times \underline{\Phi} = \underline{0}$. By Stokes' theorem, $\int_C \underline{\Phi}(\underline{r}) \cdot d\underline{r} = \iint_S (\underline{\nabla} \times \underline{\Phi}) \cdot \underline{n} dA = \iint_S \underline{0} \cdot \underline{n} \, dA = 0$.

21. Let ∂S be parametrized by $\underline{\sigma}(t)$. By Stokes' theorem, $\iint_S (\underline{\nabla} \times \underline{\Phi}) \cdot \underline{n} \, dA = \int_{\partial S} \underline{\Phi} = \int_{\underline{\sigma}(t)} \underline{\Phi} = \int \underline{\Phi}(\underline{\sigma}(t)) \cdot \underline{\sigma}'(t) dt = 0$ since $\underline{\Phi}(\underline{\sigma}(t))$ is perpendicular to $\underline{\sigma}'(t)$.

25. In symbols, the circulation of $\underline{H}$ around C is $\int_C \underline{H}(\underline{r}) \cdot d\underline{r}$. This is equal to the flux of $\underline{J}$ across S , which is $\iint_S \underline{J} \cdot \underline{n} \, dA$. Starting with $\underline{\nabla} \times \underline{H} = \underline{J}$, we take the scalar product of both sides with the outward pointing normal $\underline{n}$ to get $(\underline{\nabla} \times \underline{H}) \cdot \underline{n} = \underline{J} \cdot \underline{n}$. Integrate over S to get $\iint_S (\underline{\nabla} \times \underline{H}) \cdot \underline{n} \, dA = \iint_S \underline{J} \cdot \underline{n} \, dA$. Then, use Stokes' theorem to get $\int_C \underline{H}(\underline{r}) \cdot d\underline{r} = \iint_S \underline{J} \cdot \underline{n} \, dA$, which is Ampere's law.

29.

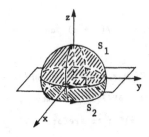

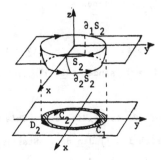

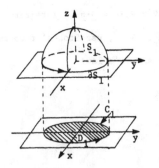

The surface S is shaded in the figure
at the left. It consists of two parts,
drawn separately in the figures below.
If $\underline{\Phi}$ is any continuously differentiable
vector field in space, then Stokes'
theorem as written in the box on p. 918
applies directly to the hemisphere S_1
to give $\int_{\partial S_1} \underline{\Phi}(\underline{r}) \cdot d\underline{r} = \iint_{S_1} (\underline{\nabla} \times \underline{\Phi}) \cdot \underline{n} dA$.

Now, we may also apply Stokes'
theorem to the surface S_2 , but a little
extra care is necessary. The orientation
of S_2 as the graph of a function on the
region D_2 (normal pointing upward) is
the opposite of its orientation as a part
of the surface S (normal pointing outward), so Stokes' theorem in this case
gives us (since the boundary of D_2 is
$C_1 - C_2$) $-\int_{\partial_1 S_2} \underline{\Phi}(\underline{r}) \cdot d\underline{r} - \int_{\partial_2 S_2} \underline{\Phi}(\underline{r}) \cdot d\underline{r} =$
$-\iint_{S_2} (\underline{\nabla} \times \underline{\Phi}) \cdot \underline{n} \, dA$. Subtracting the
second equation above from the first, and
using the facts that $S = S_1 + S_2$, $\partial S_1 =$
$-\partial_1 S_2$, and $\partial_2 S_2 = \partial S$, we obtain the result $\int_{\partial S} \underline{\Phi}(\underline{r}) \cdot d\underline{r} =$
$\iint_S (\underline{\nabla} \times \underline{\Phi}) \cdot \underline{n} \, dA$, which is Stokes' theorem for the region S .

SECTION QUIZ

1. Calculate the surface integral of $\underline{\Phi} = x\underline{i} + 2z\underline{k}$ over the surface
   S , where S is defined by $z = xy^2$ over the triangle bounded by
   $x = 0$, $y = 0$, and $y = x - 1$.

2. Let $\underline{\Phi} = 2xe^y\underline{i} + (x^2e^y + z^3)\underline{j} + 3z^2y\underline{k}$. Use Stokes' theorem to
   compute $\int_C \underline{\Phi}(\underline{r})\cdot d\underline{r}$, where C is the boundary of the semi-ellipsoid
   $x^2 + 4y^2 + 9z^2 = 25$, $z \geqslant 0$.

3. Find the scalar curl of $\underline{W}(r,s) = (s \ln r + r^2 + s)(\underline{i} + \underline{j}) + r\underline{j}$.

4. On a calm, sunny day, an old fisherman got his line caught. Unknown
   to him, the line was caught on the rusty old lock of an old sea
   fortress. Tugging with all of his might, he yanked the lock off,
   forming an odd-looking whirlpool. The velocity field of the water is
   given by $\underline{V} = (x + y)^2\underline{i} + xy\underline{j} - z\underline{k}$.

   (a) Compute curl $\underline{V}$.

   (b) The circulation is defined as $\int_C \underline{V}(x,y,z)\cdot(dx\ \underline{i} + dy\ \underline{j} + dz\ \underline{k})$.
       If C is the boundary of a surface S and $\underline{n}$ is normal to S ,
       write the circulation as a surface integral.

ANSWERS TO PREREQUISITE QUIZ

1. Counterclockwise

2. $\int_C (P\ dx + Q\ dy) = \iint_D (\partial Q/\partial x - \partial P/\partial y)\ dx\ dy$

3. $-3 \cos(1) + 13/3$

4. $6\underline{i} - 7\underline{j} - 4\underline{k}$

ANSWERS TO SECTION QUIZ

1. 1/15

2. 0

3. $s/r + 2r - \ln r$

4. (a) $-(2x + y)\underline{k}$

    (b) $-\iint_S [(2x + y)\underline{k}] \cdot \underline{n} \, dA$

18.6 Flux and the Divergence Theorem

PREREQUISITES

1. Recall how to compute a dot product (Section 13.4).

2. Recall how to compute a surface integral (Section 18.5).

PREREQUISITE QUIZ

1. Evalaute $(3\underline{i} + 2\underline{j} + \underline{k}) \cdot (-\underline{i} - 2\underline{j} + 2\underline{k})$.

2. If $\underline{\Phi} = \underline{i} + \underline{j} + \underline{k}$ is a vector field in space and S is the surface $f(x,y) = x^2 + y^2$ over the rectangle $[0,1] \times [0,2]$, what is the surface integral $\iint_S \underline{\Phi} \cdot \underline{n}$ dA ?

GOALS

1. Be able to compute the divergence of a vector field and explain its physical interpretation.

2. Be able to compute the flux of a vector field by using Gauss' divergence theorem.

STUDY HINTS

1. Flux. The flux of $\underline{V}$ across C is the volume of substance (usually air or water) crossing C per unit time as the substance moves with velocity $\underline{V}$. It is defined by $\int_C \underline{V} \cdot \underline{n}$ ds $= \int_C (Pdy - Qdx)$.

2. Divergence. This is defined by div $\underline{V} = \underline{\nabla} \cdot \underline{V}$, where $\underline{\nabla} = (\partial/\partial x)\underline{i} +$ $(\partial/\partial y)\underline{j} + (\partial/\partial z)\underline{k}$. Physically, if div $\underline{V} < 0$, a fluid is compressible, i.e., fluid is squeezing. If div $\underline{V} > 0$, the fluid is expanding or diverging. If div $\underline{V} = 0$, the fluid is incompressible or divergence free.

3. Gauss' theorem in the plane. This theorem states that flux $= \int_C \underline{V} \cdot \underline{n}$ ds $= \iint_D (\text{div } \underline{V}) dx \, dy$. Again, the conversion is made from line integral to

3.   (continued)

double integral. Counterclockwise orientation and continuous differ-

entiability remain requirements.

4.   <u>Gauss' theorem in space</u>. This theorem states that flux in space is

$\iint_{\partial W}(\underline{V} \cdot \underline{n}) dA = \iiint_{W}(\text{div } \underline{V}) dx \, dy \, dz$. Here, the conversion is made from

a double to a triple integral.

SOLUTIONS TO EVERY OTHER ODD EXERCISE

1.   The divergence of $P\underline{i} + Q\underline{j}$ is $\partial P/\partial x + \partial Q/\partial y$, so div $\underline{\Phi}$ =

$3x^2 - x^2 \cos(xy)$.

5.   The flux of $\underline{\Phi}$ across the perimeter of C is $\int_C \underline{\Phi} \cdot \underline{n} \, ds =$

$\iint_D (\text{div } \underline{\Phi}) dx \, dy$. In this case, the flux is $\int_{-1}^{1}\int_{-1}^{1}(2x - 3y^2) dx \, dy =$

$\int_{-1}^{1}(x^2 - 3xy^2)|_{x=-1}^{1} dy = \int_{-1}^{1} -6y^2 dy = -2y^3|_{-1}^{1} = -4$.

9.   div $\underline{V}$ > 0 if the fluid appears to be emerging from smaller regions.

div $\underline{V}$ < 0 if the fluid is converging.

It appears that div $\underline{V}$ > 0 at A and C ; div $\underline{V}$ < 0 at B

and D .

13. By definition, div $\underline{V}$ = $\partial P/\partial x + \partial Q/\partial y + \partial R/\partial z$, where $\underline{V} = P\underline{i} + Q\underline{j} +$

$R\underline{k}$. Here, div $\underline{V}$ = $1 + 1 + 1$ = 3 .

17. By direct calculation, we must compute $\iint_{\partial W} \underline{F} \cdot \underline{n} \, dA$ for each of the

six faces. When x = 0 , we have $0 \leqslant y \leqslant 1$, $0 \leqslant z \leqslant 1$ and $\underline{n} = -\underline{i}$.

Similarly when x = 1 , we have $0 \leqslant y \leqslant 1$, $0 \leqslant z \leqslant 1$ and $\underline{n} = \underline{i}$.

Thus, when x = 0 , $\iint_{\partial W}\underline{F} \cdot \underline{n} \, dA = \int_0^1\int_0^1 -x \, dy \, dz = \int_0^1\int_0^1 0 \, dy \, dz = 0$,

and when x = 1 , $\iint_{\partial W} \underline{F} \cdot \underline{n} \, dA = \int_0^1\int_0^1 x \, dy \, dz = \int_0^1\int_0^1 1 \, dy \, dz = 1$.

When y is constant, the faces may be described by $0 \leqslant x \leqslant 1$ and

$0 \leqslant z \leqslant 1$. $\underline{n} = \underline{j}$ when y = 1 and $\underline{n} = -\underline{j}$ when y = 0 . For y = 0 ,

17. (continued)

$\iint_{\partial W} \underline{F} \cdot \underline{n} \, dA = \int_0^1 \int_0^1 -y \, dx \, dz = \int_0^1 \int_0^1 0 \, dx \, dz = 0$. When $y = 1$,

$\iint_{\partial W} \underline{F} \cdot \underline{n} \, dA = \int_0^1 \int_0^1 y \, dx \, dz = \int_0^1 \int_0^1 1 \, dx \, dz = 1$.

When z is held constant, we have $0 \leqslant x \leqslant 1$ and $0 \leqslant y \leqslant 1$.

$\underline{n} = \underline{k}$ for $z = 1$ and $\underline{n} = -\underline{k}$ for $z = 0$. Therefore, when $z = 0$,

$\iint_{\partial W} \underline{F} \cdot \underline{n} \, dA = \int_0^1 \int_0^1 z \, dx \, dy = \int_0^1 \int_0^1 0 \, dx \, dy = 0$. When $z = 1$,

$\iint_{\partial W} \underline{F} \cdot \underline{n} \, dA = \int_0^1 \int_0^1 - z \, dx \, dy = \int_0^1 \int_0^1 (-1) \, dx \, dy = -1$. Therefore,

$\iint_{\partial W} \underline{F} \cdot \underline{n} \, dA$ over the unit cube is $0 + 1 + 0 + 1 + 0 + (-1) = 1$.

According to the divergence theorem, $\iiint_W (\text{div } \underline{F}) \, dx \, dy \, dz$ should give

the same answer. div $\underline{F} = 1 + 1 - 1 = 1$, so the triple integral is

$\int_0^1 \int_0^1 \int_0^1 dx \, dy \, dz = 1$.

21. Rearrange to get $\iiint_W (\underline{\nabla} f) \cdot \underline{\phi} \, dx \, dy \, dz + \iiint_W f \underline{\nabla} \cdot \underline{\phi} \, dx \, dy \, dz =$

$\iiint_W [(\underline{\nabla} f) \cdot \underline{\phi} + f \underline{\nabla} \cdot \underline{\phi}] \, dx \, dy \, dz = \iint_{\partial W} f \underline{\phi} \cdot \underline{n} \, dA$. By Gauss' divergence

theorem in space, this is true if $(\underline{\nabla} f) \cdot \underline{\phi} + f \underline{\nabla} \cdot \underline{\phi} = \text{div}(f \underline{\phi})$. This is

the identity proven in Example 6(a).

25. (a) By direct calculation or using Review Exercise 73(b), Chapter 15,

one may verify that $\underline{\nabla} \cdot \underline{\nabla} (p(\underline{q})/4\pi \| \underline{p} - \underline{q} \|) = 0$ whenever $\underline{q} \neq \underline{p}$;

here, differentiation is with respect to the vector variable $\underline{p}$.

For any positive ε , let Ω_ε be the region obtained from Ω by

deleting the ball B_ε of radius ε whose center is at $\underline{q}$. (We

may choose ε small enough so that the ball is contained in Ω .)

Applying the divergence theorem to $\underline{\nabla}(p(\underline{q})/4\pi \| \underline{p} - \underline{q} \|)$ on this

region, with $\underline{p}$ as the variable of integration, gives

$0 = \iint_{\partial \Omega_\varepsilon} \underline{\nabla}(\rho(\underline{q})/4\pi \| \underline{p} - \underline{q} \|) \cdot \underline{n} \, dA = \iint_{\partial \Omega} \underline{\nabla}(\rho(\underline{q})/4\pi \| \underline{p} - \underline{q} \|) \cdot \underline{n} \, dA -$

$\iint_{\partial B_\varepsilon} \underline{\nabla}(\rho(\underline{q})/4\pi \| \underline{p} - \underline{q} \|) \cdot \underline{n} \, dA$. In the last integral, $\underline{\nabla}(\rho(\underline{q})/$

$4\pi \| \underline{p} - \underline{q} \|)$ is equal to $-(\rho(\underline{q})/4\pi \varepsilon^2)\underline{n}$, because ∂B_ε is a

sphere about $\underline{q}$. Since the area of ∂B_ε is $4\pi \varepsilon^2$, we get

25. (a) (continued)

$\iint_{\partial\Omega} \underline{\nabla}(\rho(\underline{q})/4\pi\|\underline{p} - \underline{q}\|)\cdot\underline{n}\ dA = \rho(\underline{q})$. Integrating now with respect

to $\underline{q}$ over the region Ω , reordering the integrations and the

gradient operation on the left-hand side, and using the definition

of ϕ in the statement of the exercise, we obtain $\iint_{\partial\Omega} \underline{\nabla}\phi\cdot\underline{n}\ dA =$

$\iiint_{\Omega} \rho\ dx\ dy\ dz$.

(b) By the divergence theorem, $\iint_{\partial\Omega} \underline{\nabla}\phi\cdot\underline{n}\ dA = \iiint_{\Omega} \underline{\nabla}^2\phi\ dx\ dy\ dz$. Since

this formula and the result of part (a) are true for arbitrary

regions Ω , it follows that $\underline{\nabla}^2\phi = \rho$.

SECTION QUIZ

1. Let $\underline{U} = xz\underline{i} - e^y\sin z\ \underline{j} + (z - \ln x)\underline{k}$.

    (a) Calculate the divergence of $\underline{U}$.

    (b) If $\underline{U}$ is the velocity field of a fluid, is the fluid compressing,
         incompressible, or expanding at $(1,1,0)$?

2. (a) Express the flux of a vector field $\underline{B}$ across a surface S in
         terms of div $\underline{B}$.

    (b) Calculate the flux of $xy\underline{i} + 3y\underline{j} + 2z\underline{k}$ through the boundary of
         the region in space defined by $0 \leqslant x \leqslant 2$, $0 \leqslant y \leqslant 3$, $0 \leqslant z \leqslant$
         $2 + \sin x$.

3. A hat collector didn't realize it was going to rain today, so he wore
    his favorite hat when he took his afternoon stroll. A sudden cloudburst
    damaged the hat and out of anger, the mad hatter blew his top (actually,
    he punched a lot of holes through his hat). The wind is making the
    rain's velocity field $\underline{V} = 2x\underline{i} + y\underline{j} - \underline{k}$. The hat can be described
    as the cylinder $x^2 + y^2 \leqslant 1$, $0 \leqslant z \leqslant 5$. Assuming that the mad
    hatter punched enough holes in his hat to allow the rain to pass freely,
    what is the flux of the rain water passing through the hat?

ANSWERS TO PREREQUISITE QUIZ

1.   -5

2.   -4

ANSWERS TO SECTION QUIZ

1.   (a) $z - e^y \sin z + 1$

     (b) Expanding

2.   (a) $\iiint_W (\text{div } \underline{B}) \, dx \, dy \, dz$, where S is the boundary of W .

     (b) $(39/2)(5 \cos(2))$

3. 10π

18.R Review Exercises for Chapter 18

SOLUTIONS TO EVERY OTHER ODD EXERCISE

1. Let the line be parametrized by $(2t, 1 + t, 1 - 4t)$ for $0 \le t \le 1$,

    then $\int_C (xy\ dx + x \sin y\ dy) = \int_0^1 [2t(1 + t)(2\ dt) + 2t \sin(1 + t)dt] =$

    $\int_0^1 [4t + 4t^2 + 2t \sin(1 + t)]\ dt = (2t^2 + 4t^3/3)\big|_0^1 + \int_0^1 2t \sin(1 + t)dt$.

    Use integration by parts with $u = 2t$, $v = -\cos(1 + t)$ to get

    $10/3 + 2t(-\cos(1 + t))\big|_0^1 + 2\int_0^1 \cos(1 + t)dt = 10/3 - 2 \cos 2 +$

    $2 \sin(1 + t)\big|_0^1 = 10/3 - 2 \cos 2 + 2 \sin 2 - 2 \sin 1$.

5. We are integrating a gradient, so the integral is independent of path.

    $\int_C \underline{\nabla} f \cdot d\underline{r} = f(8, 2, \pi) - f(0, 1, 0)$, where $f(x, y, z) = xy^2 \cos z$. Thus,

    the integral is -32.

9. $\int_C (\sin x\ dx - \ln z\ dy + xy\ dz) = \int_1^2 [(\sin(2t + 1))(2\ dt) + (\ln(t^2))(dt/t) +$

    $(2t + 1)(\ln t)(2t\ dt)]$. $\int_1^2 2 \sin(2t + 1)dt = -\cos(2t + 1)\big|_1^2 = -\cos 5 +$

    $\cos 3$. Let $u = \ln(t^2)$, so $du = 2\ dt/t$; therefore,

    $\int_1^2 (\ln(t^2))(dt/t) = \int_0^{\ln 4} u\ du/2 = (u^2/4)\big|_0^{\ln 4} = (\ln 4)^2/4$. Integrate

    $\int_1^2 (4t^2 + 2t)(\ln t)dt$ by parts with $u = \ln t$ and $v = 4t^3/3 + t^2$ to

    get $(4t^3/3 + t^2)(\ln t)\big|_1^2 - \int_1^2 (4t^2/3 + t)dt = (44 \ln 2)/3 - (4t^3/9 +$

    $t^2/2)\big|_1^2 = (44 \ln 2)/3 + 83/18$. Thus, the answer is $-\cos 5 + \cos 3 +$

    $(\ln 4)^2/4 + (44 \ln 2)/3 + 83/18$.

13. We are integrating a gradient, so the integral is $f(\underline{\sigma}(t_2)) - f(\underline{\sigma}(t_1))$,

    where $f(x, y, z) = xze^y - z^3/(1 + y^2)$. The integral is $f(\underline{\sigma}(1)) -$

    $f(\underline{\sigma}(0)) = f(1, 0, 1/3) - f(0, -1, 1) = 1/3 - 1/27 + 1/2 = 43/54$.

17. Work is defined as the line integral $\int_C \underline{\Phi}(\underline{\sigma}(t)) \cdot \underline{\sigma}'(t)\ dt$. $\underline{\Phi}(\underline{\sigma}(t)) =$

    $e^t \underline{i} - t \exp(t^3)\underline{j}$ and $\underline{\sigma}'(t) = \underline{i} + 2t\underline{j}$, so the work done is

    $\int_0^1 (e^t - 2t^2 \exp(t^3))dt$. Let $u = t^3$ to get $\int_0^1 e^t\ dt - 2\int_0^1 (e^u/3)du =$

    $e^t\big|_0^1 - (2e^u/3)\big|_0^1 = e/3 - 1/3$.

21. A vector field, $P\underline{i} + Q\underline{j}$, is conservative if and only if it is the gradient of some function. It is the gradient of some function if $\partial P/\partial y = \partial Q/\partial x$. Here, $\partial P/\partial y = 6x^2 y$ and $\partial Q/\partial x = 6x^2 y^2$, so $\Phi(x,y)$ is not conservative.

25. $P\,dx + Q\,dy$ is an exact differential if $\partial P/\partial y = \partial Q/\partial x$. A potential function is constructed from $f(x,y) = \int_0^x a(t,0)dt + \int_0^y b(x,t)dt$ where $a(x,y) = P$ and $b(x,y) = Q$. The three variable case uses the result of Exercise 23. $\partial P/\partial y = e^y \sin x + xe^y \cos x$ and $\partial Q/\partial x = e^y \sin x + xe^y \cos x$, so this is an exact differential. An antiderivative is $f(x,y) = \int_0^x (\sin t + t \cos t)dt + \int_0^y xe^t \sin x\, dt = t \sin t \big|_0^x + xe^t \sin x \big|_0^y = x \sin x + xe^y \sin x - x \sin x + C = xe^y \sin x + C$.

29. For the differential equation, $P + Q(dy/dx) = 0$, the method of Example 1, Section 18.3 may be used if $\partial P/\partial y = \partial Q/\partial x$. $\partial P/\partial y = \cos x + 2xe^y$ and $\partial Q/\partial x = \cos x + 2xe^y$, so a solution exists. Thus, $f(x,y) = \int_0^x 2t\, dt + \int_0^y (\sin x + x^2 e^t + 2)dt = t^2 \big|_0^x + (t \sin x + x^2 e^t + 2t)\big|_0^y = x^2 + y \sin x + x^2 e^y + 2y - x^2 = y \sin x + x^2 e^y + 2y = C$. $y(\pi/2) = 0$ implies $C = \pi^2/4$, so the solution is $y \sin x + x^2 e^y + 2y = \pi^2/4$.

33. We have $\partial P/\partial y = x$ and $\partial Q/\partial x = 2x$. Since $\partial P/\partial y \neq \partial Q/\partial x$, the equation is not exact.

37. Use Green's theorem: $\int_C (P\,dx + Q\,dy) = \iint_D (\partial Q/\partial x - \partial P/\partial y)dx\,dy$.

   (a) $P = 0 = \partial P/\partial y$ and $Q = x$, so $\partial Q/\partial x = 1$; therefore $\int_C x\,dy = \iint_D (1)\,dx\,dy$, which is the area of D .

   (b) $P = y$, so $\partial P/\partial y = 1$ and $Q = 0 = \partial Q/\partial x$; therefore, $\int_C y\,dx = \iint_D (-1)\,dy\,dx$, which is the negative of the area of D .

   (c) $P = x$, so $\partial P/\partial y = 0$ and $Q = 0 = \partial Q/\partial x$; therefore, $\int_C x\,dx = \iint_D 0\,dy\,dx = 0$.

41. The curl is $\underline{\nabla} \times \underline{\Phi}$ and the divergence is $\underline{\nabla} \cdot \underline{\Phi}$. We compute $\underline{\Phi} =$ $2xz\underline{j} - 2xy\underline{k}$, so curl $\underline{\Phi} = (-2x - 2x)\underline{i} + (-2y)\underline{j} + 2z\underline{k} = 4x\underline{i} - 2y\underline{j} +$ $2z\underline{k}$; div $\underline{\Phi} = 0$.

45. The work done by a force field $\underline{F}$ going around a closed curve C is $\int_C \underline{F}(r) \cdot d\underline{r}$. By Stokes' theorem, this is the same as the surface in-tegral, $\iint_S (\underline{\nabla} \times \underline{F}) \cdot \underline{n} \, dA$, where C is the boundary of S .

49. $\iint_S \underline{F} \cdot \underline{n} \, dA = \iint_D (-Pf_x - Qf_y + R)dx \, dy$. $z = f(x,y) = (1 - x^2 - y^2)^{1/2}$ and D is the unit circle, so $f_x = -x/(1 - x^2 - y^2)^{1/2}$ and $f_y =$ $-y/(1 - x^2 - y^2)^{1/2}$. The surface integral becomes, in polar coordinates, $\int_0^{2\pi}\int_0^1 [(x^2 + 3xy^5)/z + (y^2 + 10xyz)/z + (z - xy)]r \, dr \, d\theta$, where x = r cos θ , y = r sin θ , and $z = \sqrt{1 - r^2}$. This becomes $\int_0^{2\pi}\int_0^1 [(r^2 + 3r^6\cos\theta \sin^5\theta)/\sqrt{1 - r^2} + 9r^2\cos\theta \sin\theta + \sqrt{1 - r^2}]r \, dr \, d\theta$. $\int(r^3/\sqrt{1 - r^2})dr = -\int[(1 - u)/2\sqrt{u}] du = -\sqrt{u} + u^{3/2}/3 + C = -\sqrt{1 - r^2} +$ $(1 - r^2)^{3/2}/3 + C$. $\int(r^7/\sqrt{1 - r^2})dr = -\int[(1 - u)^3/2\sqrt{u}] du = \int[(u^3 -$ $3u^2 + 3u - 1)/2\sqrt{u}] du = u^{7/2}/7 - 3u^{5/2}/5 + 3u^{3/2}/3 - u^{1/2} + C =$ $(1 - r^2)^{7/2}/7 - 3(1 - r^2)^{5/2}/5 + (1 - r^2)^{3/2} - \sqrt{1 - r^2} + C$. $\int r^3 dr =$ $r^4/4$; $\int r\sqrt{1 - r^2} = -(1 - r^2)^{3/2}/3 + C$. Therefore, after evaluating at 0 and 1 , the integral reduces to $\int_0^{2\pi}(48 \cos\theta \sin^5\theta/35 +$ 9 cos θ sin $\theta/4 - 1/3)d\theta = (8 \sin^6\theta/35 + 9 \sin^2\theta/8 - \theta/3)\big|_0^{2\pi} = -2\pi/3$.

53. Applying the divergence theorem to $\underline{\Phi}$ on a ball B_ε of radius ε about P_0 shows that the flux of $\underline{\Phi}$ through the boundary of B_ε is equal to $\iiint_{B_\varepsilon} \text{div } \underline{\Phi} \, dx \, dy \, dz$. Dividing by the volume of B_ε , letting ε approach 0 , and using the continuity of div $\underline{\Phi}$ at P_0 gives $\lim_{\varepsilon\to 0}$ [(flux of $\underline{\Phi}$ through ∂B_ε)/volume of B_ε] = (div $\underline{\Phi}$)(P_0) . We note that the roundness of B_ε is not essential. The result is true for any regions which shrink down to the point P_0 .

57. (a) Using Stokes' theorem, we let $P\,dx + Q\,dy + R\,dz$ be a differential form defined in space. Let C be the path composed of C_1 from $(0,0,0)$ to $(x,0,0)$, C_2 from $(x,0,0)$ to $(x,y,0)$ and C_3 from $(x,y,0)$ to (x,y,z). Then set $u = f(x,y,z) = \int_C (P\,dx + Q\,dy + R\,dz)$ and define $P = a(x,y,z)$, $Q = b(x,y,z)$, $R = c(x,y,z)$. We have $u = \int_0^x a(t,0,0)dt + \int_0^y b(x,t,0)dt + \int_0^z c(x,y,t)dt$, so $\partial u/\partial z = R$. Define another path through C_1 and then through C_4 from $(x,0,0)$ to $(x,0,z)$, followed by C_5 from $(x,0,z)$ to (x,y,z). Then define $u' = \int_0^x a(t,0,0)dt + \int_0^z c(x,0,t)dt + \int_0^y b(x,t,z)dt$, so $\partial u'/\partial y = Q$. We can apply Green's theorem to the rectangle D bounded by C_2, C_3, C_4, and C_5 to get $\iint_D (\partial Q/\partial y - \partial R/\partial z)dy\,dz = \int_{C_2+C_3+(-C_4)+(-C_5)} (Q\,dy + R\,dz) = \int_{C_1+C_2+C_3}(Q\,dy + R\,dz) - \int_{C_1+C_4+C_5}(Q\,dy + R\,dz) = u - u'$. $\underline{\nabla} \times \underline{\Phi} = \underline{0}$ implies $\partial R/\partial y = \partial Q/\partial z$, so the double integral over D is 0 and $u = u'$; therefore, $\partial u/\partial y = Q$.

   Now, let C be composed of C_6 from $(0,0,0)$ to $(0,y,0)$, C_7 from $(0,y,0)$ to $(x,y,0)$, and C_8 from $(x,y,0)$ to (x,y,z). Define $u'' = \int_0^y b(0,t,0)dt + \int_0^x a(t,y,0)dt + \int_0^z c(x,y,t)dt$, so $\partial u''/\partial z = R$. Then, let C be composed of C_6, C_9 from $(0,y,0)$ to $(0,y,z)$, and C_{10} from $(0,y,z)$ to (x,y,z). Define $u''' = \int_0^y b(0,t,0)dt + \int_0^z c(0,y,t)dt + \int_0^x a(t,y,z)dt$, so $\partial u'''/\partial x = P$. Apply Green's theorem to the rectangle bounded by C_1, C_8, C_9, and C_{10}, and get $\iint_D (\partial P/\partial x - \partial R/\partial z)dx\,dz = \int_{C_7+C_8+(-C_9)+(-C_{10})}(P\,dx + R\,dz) = \int_{C_6+C_7+C_8}(P\,dx + R\,dz) - \int_{C_6+C_9+C_{10}}(P\,dx + R\,dz) = u'' - u'''$. $\underline{\nabla} \times \underline{\Phi} = \underline{0}$ implies $\partial P/\partial z = \partial R/\partial x$, so the double integral over D is 0 and $u'' = u'''$;

57. (a) (continued)

therefore $u''' = u$ becuase $\partial u''/\partial z = R = \partial u/\partial z$. Consequently,

$\partial u/\partial x = P$ and $P\underline{i} + Q\underline{j} + R\underline{k} = \underline{\nabla}f = \underline{\phi}$ if $\underline{\nabla} \times \underline{\phi} = \underline{0}$.

(b)

$$\underline{\nabla} \times \underline{F} = \begin{vmatrix} \partial/\partial x & \partial/\partial y & \partial/\partial z \\ 2xyz + \sin x & x^2 z & x^2 y \\ \underline{i} & \underline{j} & \underline{k} \end{vmatrix} = (x^2 - x^2)\underline{i} - (2xy - 2xy)\underline{j} +$$

$(2xz - 2xz)\underline{k} = \underline{0}$; therefore, $\underline{F}$ is a gradient by part (a). In-

tegrating $2xyz + \sin x$ with respect to x yields $x^2 y^z - \cos x +$

$C(y,z)$. Taking the partials of this expression yields $x^2 \dot{z}$ and

$x^2 y$, the $\underline{j}$ and $\underline{k}$ components respectively. Thus $f(x,y,z) =$

$x^2 yz - \cos x + C$.

61. By the divergence theorem, $\iiint_W \text{div} (\underline{\nabla} \times \underline{\phi})dx\, dy\, dz = \iint_{\partial W}(\underline{\nabla} \times \underline{\phi}) \cdot \underline{n}\, dA$.

By definition, $\iiint_W \text{div}(\underline{\nabla} \times \underline{\phi})dx\, dy\, dz = \iiint_W \underline{\nabla} \cdot (\underline{\nabla} \times \underline{\phi})dx\, dy\, dz$, which

is zero by part (c). W is the region inside the two surfaces and $\partial W =$

S is the surface boundary, which is the union of the two surfaces. So

since $\iint_{\partial W}(\underline{\nabla} \times \underline{\phi}) \cdot \underline{n}\, dA = 0$, so does $\iint_S (\underline{\nabla} \times \underline{\phi}) \cdot \underline{n}\, dA$. Finally, by

Stokes' theorem, the line integral along the given closed curve,

$\int_C \underline{\phi}(\underline{r}) \cdot d\underline{r}$, is also zero.

TEST FOR CHAPTER 18

1. True or false.

(a) Green's theorem cannot be applied to a region described in polar
     coordinates by $1 \leqslant r \leqslant 2$ and $0 \leqslant \theta \leqslant 2\pi$.

(b) A vector field is a gradient if and only if it is conservative.

(c) The curl of a vector field is another vector field.

1. (d) The normal vector in the formula for Gauss' divergence theorem
        must be a unit vector.

    (e) Suppose that a particle moves from point A to point B along
        two different paths C_1 and C_2 . Then $\int_{C_1} \underline{\Phi} \cdot d\underline{r} = \int_{C_2} \underline{\Phi} \cdot d\underline{r}$.

2. Using standard notation as in the text, match the expression on the
    right with the expression on the left. Depending on the context, $\underline{V}$
    may be either $P\underline{i} + Q\underline{j} + R\underline{k}$ or $P\underline{i} + Q\underline{j}$.

    (a) $\iint_D (\partial P/\partial x + \partial Q/\partial y) \, dx \, dy$ (i) $\int_{\partial D} (P \, dx + Q \, dy)$

    (b) $\iint_D (\partial Q/\partial x - \partial P/\partial y) \, dx \, dy$ (ii) $\int_{\partial D} \underline{V}(\underline{r}) \cdot d\underline{r}$

    (c) $\iint_D (\underline{\nabla} \times \underline{V}) \cdot \underline{n} \, dA$ (iii) $\int_{\partial D} \underline{V} \cdot \underline{n} \, ds$

3.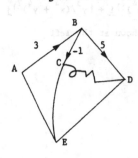

    Let $\underline{\Phi}$ be a conservative vector field in
    the plane. The line integrals of $\underline{\Phi}$ along
    the curves from A to B , from B to C ,
    and from B to D are 3 , -1 , and 5 ,
    respectively. If possible, find the line
    integrals of $\underline{\Phi}$ along the given paths:

    (a) From C to D

    (b) From D to E to A

    (c) From C to A

    (d) From E to C

4. (a) Compute the surface integral of $\underline{W} = -x\underline{i} + y\underline{j} + 3xy\underline{k}$ over the
        surface $x^2 + y^2 = z$. Let the domain of z be the region
        $0 \leqslant x \leqslant 2$, $0 \leqslant y \leqslant x^2$.

    (b) Find div $\underline{Q}$ if $\underline{Q} = (x^2 + 3xy)\underline{i} - (2x + 3y + 3e^y)\underline{j}$.

    (c) Find curl $\underline{P}$ if $\underline{P} = (x + y)\underline{i} + 2xy\underline{j} - 3(e^x - e^y)\underline{k}$.

5. Compute $\int_C (6y\ dx + 5x\ dy)$ for the following two curves, which are traversed once as shown:

(a)

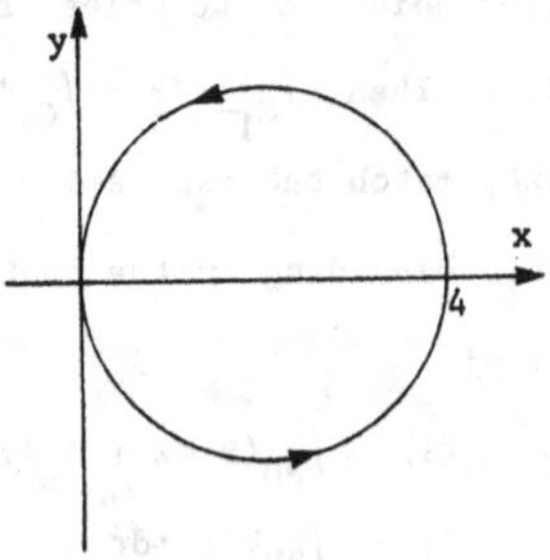

(b)

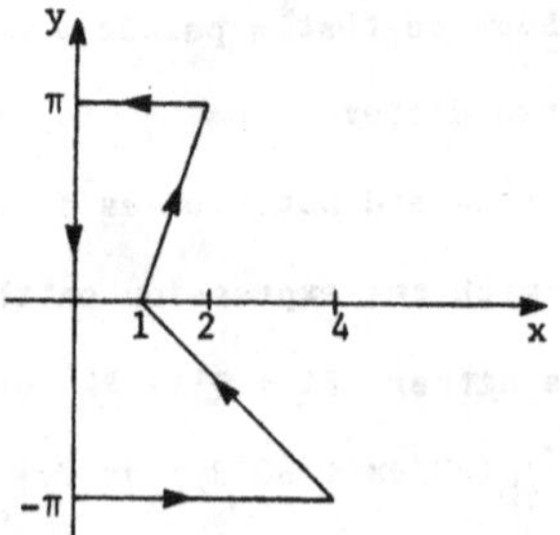

6. (a)

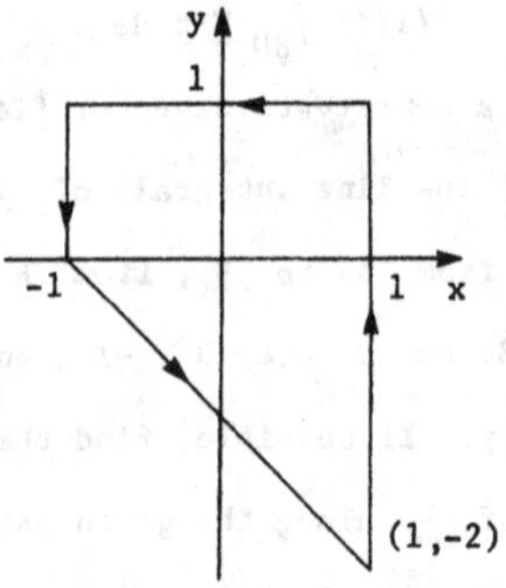

(1,-2)

Compute the line integral of $\underline{\Phi} =$ $[x^3/(x^4 + y^4)^2]\underline{i} + [y^3/(x^4 + y^4)^2]\underline{j}$ over the curve shown at the left.

(b) Show that $\partial[x^3/(x^4 + y^4)^2]/\partial y = \partial[y^3/(x^4 + y^4)^2]/\partial y$.

(c) Do your results verify or contradict any theorem? Explain.

7. Suppose that $\iint_S (\underline{\nabla} \times \underline{\Phi}) \cdot \underline{n}\ dA = 6$, where S is the hemisphere $x^2 + y^2 + z^2 = 1$, $z \geqslant 0$.

(a) What is $\iint_S (\underline{\nabla} \times \underline{\Phi}) \cdot \underline{n}\ dA$ if S is the sphere $x^2 + y^2 + z^2 = 1$?

(b) What is $\iint_S (\underline{\nabla} \times \underline{\Phi}) \cdot \underline{n}\ dA$ if S is the semi-ellipsoid $x^2 + y^2 + 4z^2 = 1$, $z \leqslant 0$?

8. Solve the exact differential equations or explain why is it not exact:

(a) $(-3x^2 y \sin(x^3 y) - 1/x) + (6 - x^3 \sin(x^3 y))dy/dx = 0$; $y(1) = \pi$.

(b) $(6 - x^3 \sin(x^3 y)) + (-3x^2 y \sin(x^3 y) - 1/x)dy/dx = 0$; $y(1) = \pi$.

9. (a) A region in space has volume $3e\pi$. Find the flux of $3z^2\underline{i} +$

$(6y - 3x)\underline{j} + (z + 5xy^2)\underline{k}$ across the boundary of the surface, if

possible. If not, explain why not.

(b)

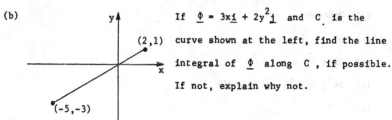

If $\underline{\Phi} = 3x\underline{i} + 2y^2\underline{j}$ and C is the

curve shown at the left, find the line

integral of $\underline{\Phi}$ along C , if possible.

If not, explain why not.

10. Your butler comes to you seeking a raise, explaining that he needs to

support his elderly mother, his wife, and his five kids. You refuse

because you know that a force vector field equal to $\underline{\nabla}f$ exists

throughout the house and is pushing him along. He claims that he works

very hard.

(a) Approximately how much work does he do if $\underline{\nabla}f = 5x\underline{i} + 4y\underline{j}$ and

he enters and leaves through the back door at $(-2,3)$?

(b) Explain your reasoning for your answer in part (a).

ANSWERS TO CHAPTER TEST

1. (a) False; the region may be divided as in Fig. 18.4.5.

(b) True

(c) True

(d) True

(e) False; only if $\underline{\Phi}$ is conservative.

2.   (a)   iii

     (b)   i

     (c)   ii

3.   (a)   6

     (b)   −8

     (c)   −2

     (d)   Unknown

4.   (a)   3884/105

     (b) $2x + 3y - 3 - 3e^y$

     (c) $3e^y \underline{i} + 3e^x \underline{j} + (2y - 1)\underline{k}$

5.   (a)   −4π

     (b)   −4π

6.   (a)   3/32

     (b)   Both equal $-8x^3 y^3/(x^4 + y^4)^3$.

     (c)   No; the cross-derivative test doesn't apply since ϕ is not

           defined at   (0,0) .

7.   (a)   0

     (b)   −6

8.   (a) $\cos(x^3 y) + 6y - \ln|x| - 6\pi + 1 = 0$

     (b)   Not exact; doesn't satisfy cross-derivative test.

9.   (a)   21eπ

     (b)   We don't know the sign of the line integral since the orientation

           isn't given.

10. (a)   0

     (b)   Work is a line integral and $\int_C \nabla f$ is independent of path. Starting

           and ending at the same point means that work is zero.

COMPREHENSIVE TEST FOR CHAPTERS 13 - 18   (Time limit: 3 hours)

1.   True or false. If false, explain why.

   (a) On a given region in the plane, a minimum of a two-variable
        function f can occur only where $\partial f/\partial x = \partial f/\partial y = 0$.

   (b) The determinant of a 2×2 or 3×3 matrix A is positive
        whenever all of the entries of A are positive.

   (c) The Gaussian integral $\int_{-\infty}^{\infty} \exp(-4x^2)dx$ is equal to $\int_{0}^{\infty} \exp(-x^2)dx$.

   (d) A circle of radius 1 , centered at the origin, has a larger
        curvature than a circle of radius 3 , centered at (7,0) .

   (e) For any two vectors $\underline{p}$ and $\underline{r}$, we have $(\underline{p} \times \underline{r}) \cdot \underline{p} = 0$.

   (f) A differentiable function of two variables, f(x,y) , has no critical
        points if either x or y is missing from the function.

   (g) The vector field $\underline{H}(x,y) = [e^{xy} + xye^{xy} - y^2\sin(xy) + 6x]\underline{i} +$
        $[x^2 e^{xy} + \cos(xy) - xy \sin(xy) - 3y^2]\underline{j}$ is a gradient.

   (h) Let $\underline{F}$ be a vector field and let C be a closed curve which is
        traversed once. Then $\int_C \underline{F} \cdot d\underline{r} = 0$.

   (i) The vector field $P\underline{i} + Q\underline{j}$ is a conservative field whenever
        $\partial P/\partial y = \partial Q/\partial x$.

   (j) The mixed partial $\partial^2 f/\partial x \partial y$ tells how fast the function f(x,y)
        is changing along the line x = y .

2.   Multiple choice.

   (a) If θ is the angle between $\underline{u}$ and $\underline{w}$, then $\|\underline{w} \times \underline{u}\|$ is:
        (i) $\|\underline{w}\| \cdot \|\underline{u}\|\cos \theta$.
        (ii) $\|\underline{w}\| \cdot \|\underline{u}\|\sin \theta$.
        (iii) $(\underline{w} \cdot \underline{u})/\|\underline{w}\| \cdot \|\underline{u}\|$.
        (iv)   None of the above.

2.   (b)   If $f(x,y,z) = g(r,\theta,z) = h(\rho,\theta,\phi)$ and W, W', $W*$ represent

the same region, then $\iiint_W f(x,y,z) \, dx \, dy \, dz$ equals:

(i) $\iiint_{W'} g(r,\theta,z) \, r \, dr \, d\theta \, dz$.

(ii) $\iiint_{W*} h(\rho,\theta,\phi) \, \rho^2 \sin^2\phi \, d\rho \, d\theta \, d\phi$.

(iii) $\iiint_{W'} g(r,\theta,z) \, r\theta \, dr \, d\theta \, dz$.

(iv) $\iiint_{W*} h(\rho,\theta,\phi) \, \rho\sin\theta \, d\rho \, d\theta \, d\phi$.

(c)   The graph of $z - x^2 = 0$ in space is:

(i)   A cylinder.

(ii)   A hyperbolic paraboloid.

(iii)   A hyperbola.

(iv)   None of the above.

(d)   If $y = f(x)$ and $z = F(x,y) = 0$, then dy/dx may be found

implicitly by the formula:

(i) $(\partial z/\partial x)/(\partial z/\partial y)$.

(ii) $(\partial z/\partial y)/(\partial z/\partial x)$.

(iii) $-(\partial z/\partial x)/(\partial z/\partial y)$.

(iv) $-(\partial z/\partial y)/(\partial z/\partial x)$.

(e)   The region described by $0 \leqslant x \leqslant 1$, $x^4 \leqslant y \leqslant \sqrt{x}$, $0 \leqslant z \leqslant$

$2 - x^2 - y^3$ is the same as the region described by:

(i) $0 \leqslant y \leqslant 1$, $y^4 \leqslant x \leqslant \sqrt{y}$, $0 \leqslant z \leqslant 2 - y^2 - x^3$.

(ii) $0 \leqslant z \leqslant 2$, $x^4 \leqslant y \leqslant \sqrt{x}$, $0 \leqslant x \leqslant 2 - z^2 - y^3$.

(iii) $0 \leqslant r \leqslant 1/\cos\theta$, $r^3\cos^4\theta \leqslant \sin\theta \leqslant r^{-1/2}\sqrt{\cos\theta}$,

         $0 \leqslant z \leqslant 2 - r^2\cos^2\theta - r^3\sin^3\theta$.

(iv) $0 \leqslant y \leqslant 1$, $y^2 \leqslant x \leqslant \sqrt[4]{y}$, $0 \leqslant z \leqslant 2 - x^2 - y^3$.

3.   Short Answers.

(a) To apply Green's theorem on a region, what condition must hold on
     the region's boundary and what condition must hold for the
     integrand?

(b) Define a function of three variables.

(c) What conditions must be satisfied by partial derivatives to
     guarantee that $f(x_0, y_0)$ is a local maximum.

(d) Find the most general function $f(x,y)$ such that $\partial^2 f / \partial x \partial y =$
     $3xy - y$.

(e) Define a local minimum of $f(x,y)$.

4.

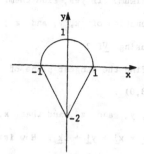

The region at the left is bounded by a
semicircle of radius 1 and two straight
lines.

(a)   Write the volume between this region
      and the graph of $f(x,y) = x^2 + y^2$
      as a double integral.

(b)   Compute the volume. [Hint:

$\int x^2 \sqrt{a^2 - x^2}\, dx = (x/8)(2x^2 - a^2)\sqrt{a^2 - x^2} + (a^4/8)\sin^{-1}(x/a)$,

if $a > 0$.]

(c) Write the volume as a triple integral.

(d) Compute the surface integral of $\underline{\Phi} = (1/3)\underline{i} + y^2\underline{j} + 3y^4\underline{k}$ over
     the surface $f(x,y) = x^3 + y^3 - y$. Let the domain be the
     region sketched above.

5. (a) Sketch the oriented curve C: $(t, t, \sin t\pi)$, $-1 \leqslant t \leqslant 3/2$.

   (b) What is the acceleration vector along C ?

   (c) If $\underline{\Omega} = xy\underline{i} - yz\underline{j} + \underline{k}$, compute the line integral of $\underline{\Omega}$ along C .

   (d) If $\underline{\Phi} = \underline{\nabla}f$, where $f(x,y,z) = xy^2 + 5e^z - 3 \cos z\pi + 4$, what is
   the line integral of $\underline{\Phi}$ along C ?

6. Let $f(x,y) = (x^2 + 3y^2)\exp(1 - x^2 - y^2)$.

   (a) Classify all critical points of f .

   (b) Does f have a global minimum or maximum? If yes, find them.

   (c) Suppose that the domain of f is $[-2,2] \times [-1/2,1/2]$. Now,
   does f have a global minimum or maximum? If yes, find them.

7. (a) If $\underline{r} = x\underline{i} + y\underline{j} + z\underline{k}$ and f is a function of x, y, and z ,
   how is the tangent plane defined by using $\underline{\nabla}f$?

   (b) Use the definition from part (a) to find the tangent plane of
   $x^2 + y^2 + z^2 = 10$ at the point $(1,3,0)$.

   (c) Suppose that g is a function of x, y, and z , and that x, y,
   and z are functions of t . Let $\underline{r} = x\underline{i} + y\underline{j} + z\underline{k}$. How is
   dg/dt defined in terms of the gradient?

   (d) Use the definition from part (c) to compute dg/dt if $g(x,y,z) =$
   $xy + z^2$, $x = e^t$, $y = t + \sin t$, and $z = 1/t$.

8. (a) Write $\int_C (x^3y^2 \, dx + x^3\cos(y^2) \, dy)$ as a double integral over D ,
   where C is the boundary (oriented counterclockwise) of D .

   (b) Compute $\iint_S (\underline{\nabla} \times \underline{B}) \cdot \underline{n} \, dA$, where $\underline{B} = xy\underline{i} + y\underline{j} + z\underline{k}$ and S is
   the part of the unit sphere lying above the plane $z = 1/2$.

   (c) Use Green's theorem to find the area of the region between $y =$
   $x^2 - 1$ and the parametrized curve, $x = \cos t$, $y = 2 \sin t$,
   $0 \leqslant t \leqslant \pi$.

9. (a) If $y = k(m,n)$, state the definition of $\partial y / \partial m$.

(b) Compute $\partial t / \partial p$ and $\partial t / \partial s$ if $t(m,p,r,s,u,v,z) = \exp(\cos rv) - \sin(\ln z) + mu/r - 1/p$.

(c) Let $f(x,y,z) = y - 2xz^2$. Compute the directional derivative of f in the direction of $\underline{i} - 2\underline{j} + 2\underline{k}$ at $(x,y,z) = (1,1,1)$.

(d) Find the direction in which the directional derivative of $f(x,y,z) = y - 2xz^2$ at $(x,y,z) = (1,1,1)$ is maximized.

10. Miscellaneous problems.

(a) Find the flux of $\underline{K} = (3x + y + z)\underline{i} + (x + 5y + z^2)\underline{j} + (x - y^3 + 6z)\underline{k}$ across the parallelepiped spanned by the vectors $\underline{i} + \underline{j} + \underline{k}$, $2\underline{i} - \underline{k}$, and $3\underline{i} + \underline{j} - 2\underline{k}$.

(b) What is curl $\underline{K}$ for $\underline{K}$ as in part (a)?

(c) Sketch the graph of $2^2 + (-1)^2 + z^2 = 25$.

(d) Write down $\partial(x,y,z)/\partial(\rho,\theta,\phi)$ for x, y, and z defined by spherical coordinates.

11. Extra Credit. Choose the best answer.

Calculus is _____ .

(a) a lot more enjoyable than going through a torture chamber.

(b) a fun hobby for A students.

(c) a combination of two four-letter words.

(d) worse than an evening with the in-laws.

ANSWERS TO COMPREHENSIVE TEST

1.   (a) False; the minimum may occur on the boundary of the region.

   (b) False; let $A = \begin{bmatrix} 0 & 1 \\ 1 & 0 \end{bmatrix}$.

   (c) True

   (d) True

   (e) True

   (f) True

   (g) True

   (h) False; $\underline{F}$ must be conservative to guarantee that $\int_0 \underline{F} \cdot d\underline{r} = 0$.

   (i) False; $P\underline{i} + Q\underline{j}$ must be defined throughout the domain.

   (j) False; the directional derivative gives that information.

2.   (a) ii

   (b) i

   (c) i

   (d) iii

   (e) iv

3.   (a) The boundary must be traversed counterclockwise; the partials
         must be continuous on the region.

   (b) A rule which assigns a single value to each point (x,y,z) in the
         domain.

   (c) At (x_0, y_0) , $f_x = f_y = 0$; $f_{xx} < 0$; and $f_{xx} f_{yy} - f_{xy}^2 < 0$.

   (d) $3x^2 y^2 /4 - xy^2 /2 + g(x) + h(y) + \text{constant}$, where g is a function
         of x only and h is a function of y only.

   (e) (x_0, y_0) is a local minimum point of f if there is a disk about
         (x_0, y_0) such that $f(x,y) \geqslant f(x_0, y_0)$ for all (x,y) in the disk.

4. (a) By symmetry, $V = 2\int_0^1 \int_{2x-2}^{\sqrt{1-x^2}} (x^2 + y^2)\, dy\, dx$.

   (b) $\pi/4 + 5/3$

   (c) $V = \int_0^1 \int_{2x-2}^{\sqrt{1-x^2}} \int_0^{x^2+y^2} dz\, dy\, dx$

   (d) $-\pi/4 - 5/3$

5. (a)

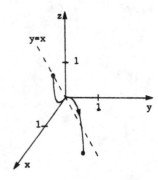

   (b) $(0,0,-\pi^2 \sin \pi t)$, $-1 \le t \le 3/2$

   (c) $35/24 + 1/\pi^2$

   (d) $11/8 + 5e^{-1}$

6. (a) $(0,0)$ = local minimum; $(0,\pm 1)$ = local maximum; $(\pm 1, 0)$ = saddle point

   (b) Minimum: $(0,0,0)$; maximum: $(0,\pm 1,3)$

   (c) Minimum: $(0,0,0)$; maximum: $(\pm 1/2, \pm 1/2, \sqrt{e})$

7. (a) $\underline{\nabla} f(\underline{r}_0) \cdot (\underline{r} - \underline{r}_0)$

   (b) $x + 3y = 10$

   (c) $\underline{\nabla} f(\underline{r}) \cdot (d\underline{r}/dt)$

   (d) $(1 + t + \sin t + \cos t)e^t - 2/t^3$

8. (a) $\iint_D (3x^2 \cos(y^2) - 2x^3 y)\, dx\, dy$

   (b) 0

   (c) $4/3 + \pi$

9. (a) $\lim\limits_{\Delta m \to 0} \{[\, k(m + \Delta m,n) - k(m,n)]\, / \Delta m\}$

    (b) $\partial t / \partial p = 1/p^2$; $\partial t / \partial s = 0$

    (c) -4

    (d) $-2\underline{i} + \underline{j} - 4\underline{k}$

10. (a) 52

    (b) $(-3y^2 - 2z)\underline{i}$

    (c)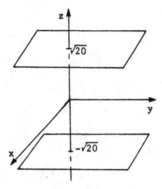

    (d) $\begin{bmatrix} \sin\phi\cos\theta & -\rho\sin\phi\sin\theta & \rho\cos\phi\cos\theta \\ \sin\phi\sin\theta & \rho\sin\phi\cos\theta & \rho\cos\phi\sin\theta \\ \cos\phi & 0 & -\rho\sin\phi \end{bmatrix}$